OXFORD READINGS IN FEMINISM

FEMINISM AND SCIENCE

OXFORD READINGS IN FEMINISM

Feminism and Science

Edited by
Evelyn Fox Keller
and **Helen E. Longino**

Oxford · New York

OXFORD UNIVERSITY PRESS

1996

Oxford University Press, Walton Street, Oxford OX2 6DP
Oxford New York
Athens Auckland Bangkok Bombay
Calcutta Cape Town Dar es Salaam Delhi
Florence Hong Kong Istanbul Karachi
Kuala Lumpur Madras Madrid Melbourne
Mexico City Nairobi Paris Singapore
Taipei Tokyo Toronto
and associated companies in
Berlin Ibadan

Oxford is a trade mark of Oxford University Press

Published in the United States
by Oxford University Press Inc., New York

British Library Cataloguing in Publication Data
Data available

Library of Congress Cataloging in Publication Data
Feminism and science / edited by Evelyn Fox Keller and Helen E. Longino.
p.   cm.   — (Oxford readings in feminism)
Includes bibliographical references (p.   ) and index.
1. Science—Social aspects. 2. Science—Philosophy.
3. Feminist theory. 4. Women in science.
I. Keller, Evelyn Fox. 1936–   . II. Longino, Helen E. III. Series.
0175.5.F455 1995   306.4'5—dc20   95-45299
ISBN 0–19–875146–X (Pbk.)
ISBN 0–19–875145–1

10 9 8 7 6 5 4 3 2 1

Photo of Helen E. Longino taken
by Tom King

Typeset by Hope Services (Abingdon) Ltd.
Printed in Great Britain by
Bookcraft (Bath) Ltd
Midsomer Norton, Avon

Contents

Notes on Contributors

CAROL COHN teaches women's studies and sociology at Bowdoin College.

RUTH DOELL, Professor Emerita of Life Sciences at San Francisco State University, is author of numerous papers in biochemistry, immunology, and cancer, as well as of papers on science and society.

CHRISTINE GRONTKOWSKI is Professor of Philosophy and Dean of the College of Liberal Arts and Sciences at Alfred University. Her Ph.D. in Philosophy of Science is from Fordham University and she has taught at Fordham, Vassar College, the State University of New York at Purchase, and Alfred University.

DONNA HARAWAY is a Professor in the History of Consciousness Board at the University of California at Santa Cruz. She is author of *Primate Visions: Gender, Race, and Nature in the World of Modern Science* (Routledge, 1989), and *Simians, Cyborgs, and Women: The Reinvention of Nature* (Routledge and Free Association Books, 1991).

SANDRA HARDING is Professor of Philosophy and Women's Studies at the University of Delaware and the University of California, Los Angeles. She is most recently author of *Whose Science? Whose Knowledge?* (Cornell University Press, 1991) and editor of *The 'Racial' Economy of Science: Toward a Democratic Future* (Indiana University Press, 1993).

EVELYN FOX KELLER is Professor of History and Philosophy of Science in the Program in Science, Technology, and Society at the Massachusetts Institute of Technology. Her most recent book, *Refiguring Life: Metaphors of Twentieth-Century Biology* has been published by Columbia University Press, 1995.

ELISABETH A. LLOYD is Associate Professor of Philosophy at the University of California, Berkeley. She is author of *The Structure and Confirmation of Evolutionary Models* and co-editor with Evelyn Fox Keller of *Keywords in Evolutionary Biology* (Harvard University Press, 1992).

GENEVIEVE LLOYD is Professor of Philosophy at the University of New South Wales, Sydney, Australia, and author most recently of *Being in Time: Selves and Narrators in Philosophy and Literature* (Routledge, 1993) and *Part of Nature: Self-Knowledge in Spinoza's Ethics* (Cornell University Press, 1994).

HELEN E. LONGINO teaches women's studies and philosophy at the University of Minnesota. She is author of *Science as Social Knowledge: Values and Objectivity in Scientific Inquiry* (Princeton University Press, 1990).

EMILY MARTIN is Professor of Anthropology at Princeton University.

Author of *The Woman in the Body: A Cultural Analysis of Reproduction* (Beacon, 1987), her latest research is described in *Flexible Bodies: Tracking Immunity in America from the Days of Polio to the Age of AID* (Beacon, 1994).

NAOMI SCHEMAN is a Professor of Philosophy and Women's Studies at the University of Minnesota. She is the author of *Engenderings: Constructions of Knowledge, Authority, and Privilege* (Routledge, 1993).

LONDA SCHIEBINGER is Professor of History and Women's Studies at the Pennsylvania State University. She is author of *The Mind has No Sex? Women in the Origins of Modern Science* (Harvard University Press, 1989) and of *Nature's Body: Gender in the Making of Modern Science* (Beacon Press, 1993).

DOROTHY E. SMITH is Professor of Sociology and Head of the Centre for Women's Studies in Education, Ontario Institute for Studies in Education. Her most recent books are *The Conceptual Practices of Power: A Feminist Sociology of Knowledge* (Northeastern University Press, 1990) and *Texts, Facts, and Femininity: Exploring the Relations of Ruling* (Routledge, 1990).

NANCY LEYS STEPAN is Professor of History at Columbia University, where she teaches the history of science and medicine. Author of *The Idea of Race in Science: Great Britain 1800–1960* (Archon Books, 1982), she is currently working on visuality and the natural world.

MARY TILES, formerly Secretary to the Royal Institute of Philosophy and lecturer at Oxford and Cambridge universities, is presently Professor of Philosophy at the University of Hawaii at Manoa. Her books include *Bachelard: Science and Objectivity* (Cambridge University Press, 1984), and *Mathematics and the Image of Reason* (Routledge, 1991).

Introduction

Evelyn Fox Keller and Helen E. Longino

The natural sciences have assumed a position of unparalleled authority in twentieth-century Western intellectual life. But what kind of knowledge do the sciences provide us and what is the basis of their cognitive authority? These questions have grown more pressing as feminist and environmental movements have implicated scientific research and science-based technologies in the continuing subordination of women and in the degradation of the environment. Are the sciences neutral with respect to social issues and social values, their harm or benefit arising solely from the uses to which knowledge, once obtained, is put? Or do they develop in more intimate interaction with their social and cultural contexts, reflecting particular social and cultural values?

Until the 1960s the dominant view of the sciences was that scientific knowledge consisted of logical reasoning applied to observational and experimental data acquired by value-neutral and context-independent methods. It was also widely believed that the application of scientific methods in developing knowledge of nature resulted (or would in time result) in a single, unified, account of an objective and determinate world. In the 1960s, however, the work of several historians of science and historically minded philosophers of science—Thomas S. Kuhn, Paul Feyerabend, and N. Russell Hanson—decisively challenged this vision. Scientific observation, they argued, is never innocent, but always and inevitably influenced by theoretical commitments. Observation, in a word, is theory-laden. Moreover, the development of scientific knowledge cannot be understood as a matter of accumulation, the addition of more detail or theoretical sophistication to a stable base. Stability itself is temporary, subject to periodic rupture in the course of what Kuhn termed scientific revolutions. While these arguments were not and are still not accepted by everyone, they have inspired a wealth of studies in history and social

1

studies of science detailing the ways in which the intellectual and social context of scientific research can significantly affect its course. And much philosophy of science since the 1960s has consisted of attempts to rethink and reformulate claims of the objectivity of scientific method in response to the challenges posed to the more traditional philosophical views.

The 1970s brought the advent of contemporary feminist theory—a sustained inquiry into the ways in which the conventional disciplines had been shaped by their historic exclusion (or misrepresentation) of the lives and experiences of women. These efforts initially focused only on disciplines remote from the history and philosophy of science, but they soon made their impact felt in all areas of inquiry. The critical theoretical concept was that of 'gender', introduced as a way of distinguishing the social constitution of masculine and feminine from the biological categories of male and female. Very soon, however, feminists began to recognize this concept as an analytic tool for studying the impact of gender ideology on the structure of social and intellectual worlds extending far beyond the minds and bodies of individual men and women. Using gender as a lens, or as an analytic tool for examining the implications of coding some spheres as masculine and others feminine, scholarship in the social sciences and humanities soon produced dramatic and persuasive reinterpretations of many familiar and well-studied matters (including institutions, practices, cultural artefacts, historic periodization, literary standards, etc.).

The earliest contemporary feminist scholarship on the natural sciences tended to focus on the barriers aspiring women scientists have faced in the past (and continued to face in the present), as well as on the recuperation of those women scientists of the past who had been erased from the historical record. But the concept of gender opened an entirely new window on the nature of scientific inquiry. Combining the insights of feminist theory with those of contemporaneous developments in the history, philosophy, and sociology of science has added a new dimension of analysis; it has raised important yet heretofore unasked questions about the content and practice of the natural sciences, about the forms of interaction with the rest of the natural world that scientists have historically cultivated, and about the goals that have traditionally been idealized in the natural sciences. Feminist critics of science have argued that modern science evolved out of a conceptual structuring of the world—e.g. of mind and nature—that incorporated particular and historically specific ideologies of gender.

Furthermore, they argue, the effects of such incorporations continue to be evident in the cultures and practices of contemporary natural science.

In this volume, we have collected a number of readings to reflect the diversity and strengths of work in this area. In Part One, we include three early (one might even say, preliminary) statements of the range of problems in science studies that feminist theory brings into view, written from three different disciplinary perspectives. Dorothy Smith writes as a sociologist, Evelyn Keller as a scientist, and Genevieve Lloyd as a historian of philosophy. Yet the common denominator underlying both their concerns and their perspectives is striking. That common denominator was provided by the women's movement of the 1970s, and its insistence that sexual equity in any domain, political or intellectual, depended on first incorporating into these domains the experiences and perspectives of women's social realities. One might say that feminist theory emerged out of the question that quickly follows: How does the world (political or intellectual) change with the incorporation of the experiences and perspectives of women? Thus, in a particularly early contribution to feminism and science, and speaking for sociology, Smith observes that access to the traditionally discounted, or repressed, realities of women might even

bring us to ask first how a sociology might look if it began from the point of view of women's traditional place in it and what happens to a sociology which attempts to deal seriously with that. Following this line of thought, I have found, has consequences larger than they seem at first. (p. 17)

Among other consequences, it leads Smith to a general analysis of the dependence of knowledge in the social sciences on subject position, and to a clear and forceful presentation of what has since come to be labelled as 'standpoint theory'.

A similar logic leads Keller not to 'standpoint theory', but to a challenge of the supposed universality of scientific norms. Starting from a critique of 'unfair employment practices' responsible for the underrepresentation of women in science, she proceeds, first, to critiques of identifiable androcentric bias in the (usually biological or social) sciences, and finally, to 'the possibility of extending the feminist critique into the foundations of scientific thought'. Her aim is not a 'different science', but a liberated one, open to 'potentialities lying latent in the scientific project', but excluded or repressed by its 'masculinist' history.

3

Genevieve Lloyd's project is a more historical one. It derives from much the same impetus as Smith's and Keller's projects, but, as a philosopher, Lloyd is more attentive to the costs of a masculinist history to philosophy than to sociology or to science. The selection included here is excerpted from her book, *The Man of Reason* (1984), in which she writes:

The maleness of the Man of Reason, I will try to show, is no superficial linguistic bias. It lies deep in our philosophical tradition. This is not to say that women have their own truth, or that there are distinctively female criteria for reasonable belief. It is, however, to make a claim which is no less a scandal to the pretensions of Reason. (p. ix)

Like her project, the claim Lloyd makes is also a historical one— not about the nature of reason, but about what has historically been construed by philosophers as reason, namely, the thing we call 'Reason'. And about this, she concludes: 'Our trust in a Reason that knows no sex has . . . been largely self-deceiving' (p. ix). In the chapter included here, Lloyd also draws attention to the correlative metaphoric identification of nature with the female. This has not been a constant feature of thought but has roots in an antecedent identification in the classical era of matter and the female. Both of these identifications reserve the position of knower, and hence Reason, to the male.

Each of the papers of this part may be thought of as an attempt to sketch out a terrain for deeper and more extensive exploration. As the next sections demonstrate, they collectively identify a number of the questions that have since been pursued by feminist scholars in greater depth.

Part Two includes readings that critically explore problems of representing sex and gender, or, as they are sometimes called, issues in 'the science of gender'. These issues have a considerably longer history than many others discussed in this volume. At least since the eighteenth century, feminists have contested the ways in which scientists typically represented female animals (including women), gender relations, and sexual reproduction. Judith Drake in the eighteenth century and Antoinette Blackwell and other feminist intellectuals in the nineteenth are forerunners to the contemporary resistance to masculine bias in the science of gender (Cohen 1994; Russett 1989).

In the current wave of feminist scholarship, women scientists have led the way in documenting the effects of gender bias in the collection, organization, and interpretation of data on sex differences from current research in the behavioural, biological, and biobehavioural sciences. Emboldened by feminism, these critics documented the use of exclusively male subjects in both experimental and clinical biomedical research, the selection of male-dominant animal populations as model species in ethology, and the focus on male dominance and activity and concomitant invisibility of females in research protocols. (Bleier 1984; Fausto-Sterling 1985) Historians, philosophers, and sociologists/anthropologists took up this work, using the tools of their disciplines to clarify the resultant debates and develop them as resources for challenging mainstream views in their disciplines about scientific objectivity and value-neutrality.

Donna Haraway, a historian of twentieth-century primatology, has integrated historical methods with the critical perspectives of twentieth-century social theory to analyse the intertwining of agendas of primate research with those of social control. She sets particular research programmes in their larger intellectual and sociopolitical contexts, analysing professional networks within and across fields to lay bare the social aims and ideological commitments threaded through primatological projects. In the essay included here, Haraway traces shifts in theories of the biosocial conditions of production and reproduction, focusing on debates about hominid origins and about the physiological bases of primate social organization. She emphasizes that contests over 'male-centred' versus 'female-centred' accounts must be understood in the context of the political, economic, and cultural logics that give such divisions by gender their meanings. Haraway notes that all of the research programmes she discusses follow rigorous methodological protocols and that none can be faulted as bad science, but she finds hope for the future in the articulation of an alternate past.

Helen Longino and Ruth Doell, writing as a cross-disciplinary team, submit a range of biological studies of sex differences to philosophical analysis. By doing so, they are able to demonstrate the permeability of inquiry to culturally based assumptions. Their concern is to show the different ways in which the structure of inquiry permits the expression of ideology in the content of research. It's not a matter of the wilful imposition of stereotypes against the evidence of the senses, but a matter of the organization and interpretation of data in the context of background assumptions. Focusing on research performed in the 1960s and 1970s, they are concerned particularly to show how behavioural neuroendocrinological

5

research was differently organized by gender ideology than was the human evolution research discussed by Haraway. These differences show that different strategies of feminist intervention are warranted by different research contexts.

Philosopher of biology Elisabeth Lloyd pursues one such strategy in her essay on the treatment of female sexuality. She finds that an insistence on tying sexuality to reproduction in theorizing about the evolution of sexuality overrides data concerning actual sexual behaviour. Lloyd attributes this discordance of theory and data to the integration of adaptationism (the view that any trait is to be explained in terms of its adaptive or functional value) with the long-standing cultural definition of women in terms of their reproductive function.

Emily Martin pursues the theme of gender stereotypes, focusing on their role in the biology of sexual reproduction as described in college and medical textbooks. The female gamete, the egg, is repeatedly treated as a passive object towards which the energetic sperm propels itself and on which it acts. Martin argues that the persistence of these stereotypes not only demonstrates the play of gender ideology in the scientific description of natural processes, but, given the cognitive authority of the sciences, serves to reinforce them in the larger culture. Indeed, her primary concern is with their role in that culture; the main concern in this volume, however, is with the effect of gendered imagery on the sciences themselves. Other scholars have pursued similar analyses—using even the very examples Martin cites—to show how such imagery can influence the course of experimental research by lending certain kinds of processes an aura of unthinkability (see, for example, Keller 1995).

Sex and gender, even in the sciences, may be thought of as essentially contested or contestable subjects. Indeed, the feminist critique is often dismissed as limited to the correction of bias about such topics—regrettable, but understandable. The papers of Haraway and Longino and Doell show that this is an incorrect assessment. Research bearing on sex and gender is not differentiated in its social and logical structures from other research. Thus, not only do the papers included here dissect particular areas of research, they offer models of analysis that can be applied to any research programme. The challenge to extend the feminist critique to non-gendered subject matter has been taken up primarily by scholars focusing on language and metaphor. This large subject, the role of language in shaping research agendas beyond the human sciences, is the topic of the next section.

The common concern of the papers in Part Three is with a variant of Hacking's question (1975): How does language matter to science? More specifically, all of the essays included here are focused on the role of images, metaphors, and analogies of gender in the construction of scientific representations, even of non-human or non-gendered subjects. Nancy Stepan draws on the particular use of analogies of race and gender in the human sciences for her main example, but her interest is in an analysis of 'the role of metaphors, and of the analogies they mediate, in scientific theory' (p. 121) that would hold more generally. She uses a particular example to illustrate how metaphors guide the construction of similarities and differences—i.e. of our very categories of analysis. But the mechanism illuminated by her example, she notes, is how metaphors always work. As she concludes,

We need a critical theory of metaphor in science in order to expose the metaphors by which we learn to view the world scientifically, not because these metaphors are necessarily 'wrong', but because they are so powerful. (p. 133)

The subject of Londa Schiebinger's essay shifts the focus from the construction of human taxonomies to the construction of animal taxonomies. Mammals, she observes, are grouped together and demarcated from other animals by a distinctly female function—the presence of mammary glands defines the mammal as the genus of nurturing mothers. By contrast, humans (*Homo sapiens*) are grouped together, and distinguished from other mammals, by their intelligence. The effect of this taxonomic structure on perceptions of gender is obvious (women are mammals; men are human). And so is the converse. Schiebinger suggests that the asymmetric gendering of *Mammalia* and *Homo sapiens* helped secure acceptance of the controversial taxonomic classification of humans with other animals. As befits a historian, however, Schiebinger's primary concern is with the particular social and political context of the introduction, by Linnaeus, of the term *Mammalia*, and with its acceptance at that particular time and place by his larger community.

In the next essay, disciplinary differences surface once again. Here, Keller aims to make the case for the force of language, and more specifically for the inhibitory costs of gendered metaphors, to working scientists. Her intended audience is the community of theoretical and mathematical evolutionary biologists, and her goal is to use her analysis of language to reveal problems in this area of research that had otherwise remained hidden or overlooked. She relies for her main examples on the language of competition and reproduction, but, as with Stepan, her message is a more general one. She concludes that the inevitable slippage between ordinary and technical meanings of such terms provides an ever-ready vehicle for the importation of cultural and political values into ostensibly purely technical arenas.

Finally, Carol Cohn's essay extends the discussion beyond the biological sciences. Not quite the physical sciences, her subject is the language of defence intellectuals. But again, her analysis is more general. Her principal finding, developed in the course of a fellowship at one of the major centres of defence policy analysis in the United States, is of the force with which language constrains our vision of what is possible. Metaphors of sexuality and domesticity serve to familiarize the grotesque, to make thinkable the unthinkable. At the same time, they nullify or 'sissify' the notion, say, that nuclear annihilation is unthinkable just because it is a priori unacceptable. Furthermore, and perhaps most importantly, she documents how readily such metaphors (and such a discourse), initially so startling, are adapted to:

Within a few weeks, what had once been remarkable became unnoticeable. . . . I had not only learned to speak a language: I had started to think in it. (p. 182)

Here, Cohn articulates the central theme of this section: what it means to think, and accordingly to act, in a particular language.

The papers in the first part provide clear evidence of the historically close relation between cultural ideals of masculinity and conventional conceptions of knowledge and reason. Those in the second part demonstrate the inadequacy of the view that masculinist research on sex and gender can be simply dismissed as 'bad science'. And those in the third part track the force of gender ideology in science through an examination of the role of metaphor (e.g. metaphors of gender) in scientific practice. All three lines of argument raise new problems for the theory of knowledge, and hence they require, and in fact begin to forge, new approaches to this subject.

The problems are familiar. Is it possible to give an adequate account of the role of the subject (i.e. the knower) in constituting the object of knowledge without losing sight of the role of the object? How does one challenge the cognitive authority of science while simultaneously maintaining that there really are more and less adequate accounts of natural processes? What other standards are to be invoked by which to measure such adequacy? And finally, and most familiarly, how does one reject objectivism without falling into the trap of relativism?

Perhaps they are too familiar. All of these problems are themselves inherited from an ancient tradition of oppositions and dichotomies. By contrast, virtually all the essays collected in this volume begin by challenging the familiar dichotomies between subject and object, transcendent and immanent, mind and body. They seek, instead, as an alternative to the traditionally idealized 'view from nowhere', a philosophy of embodied and socially, temporally, and spatially situated knowledge. Their criticism is not so much that science has failed in its traditional aspirations, but that these aspirations themselves may be misguided. The readings in this, the final part of the book, return to the themes of the first part, and trace the history of concepts and metaphors that have linked cognition, rationality, and masculinity. But they also go beyond the early critiques in attempting to chart positive alternatives to traditional conceptions of subjectivity and objectivity and to consider the role these positive alternatives might have in scientific inquiry. Though differing significantly from one another, these efforts have in common an emphasis on pluralism, community, and reflexivity in the development of their alternative conceptions.

'The Mind's Eye', jointly authored by Evelyn Fox Keller and Christine Grontkowski, focuses on the history and consequences of taking vision as a metaphor for knowledge. Developing themes adumbrated in Genevieve Lloyd's discussion of Plato in our first part, they show how use of such a metaphor has permitted particular theories of light and vision to shape our concepts of knowledge, while simultaneously promoting the epistemological ideals of detachment and transcendence. They suggest that other senses—touch, for example—have historically been regarded as inappropriate models just because they are more obviously entangling, more conspicuously embodied. Yet the question is worth asking: What difference might the metaphoric use of other sensual modalities have made to our conceptions of knowledge? Furthermore, even the visual experience, they argue, is double-edged—establishing not simply distance and objectification (as recent feminist theorists of the 'male gaze' have argued), but also communion. Keller and Grontkowski argue that the historical neglect of this dimension of vision is part and parcel of the equally historical repression of the erotic dimensions of knowledge.

In her essay, Naomi Scheman, a philosopher, employs a psycho-analytic perspective to examine that particular concern of René Descartes's that has so influenced epistemologists' definitions of knowledge, namely his preoccupation with the search for absolute certainty. Scheman suggests that both Descartes's worries and his solutions to them have recognizable parallels in the psychic condition diagnosed by Freud as paranoia. Paranoia is described by Freud as a process of splitting and alienation, whereby something that had been recognized as part of the self is split off and reattached to something or someone other than oneself. The motive for such splitting is a perceived incompatibility of what is thereby rejected with the developing ego. Just so, Cartesian method requires splitting the body from the 'real self', which Descartes analysed as a non-material substance, different in kind from the material world. Scheman argues that, in our actual circumstances of social inequality, that which was so rejected, namely the body, is reattached to the persons of all those excluded from social privilege—namely, women, members of ethnic and racial minorities, the working class, sexual minorities, all of whom are identified almost exclusively with bodily traits or bodily functions. She argues that reintegration of mind and body would require new conceptions of subjectivity, which, in turn, would enable new epistemic relationships.

The possibility of such new relationships is at the heart of Mary Tiles' essay. Like many of the authors in this volume, hers is a call

for the development of science with a human face, science which aims more at co-operating with than at conquering Nature, which learns more by conversing or conducting a dialogue with Nature than by putting it on the rack, . . . a science of Venus rather than a science of Mars. (p. 221)

Careful to separate 'normative' claims about how science should proceed from those she calls 'evaluative' (about how successful science has been), her particular focus is on the ways in which the 'logic of domination'—i.e. the historic interest of modern science in prediction and control—has shaped criteria for success. Recognition of the particularity of such interests undermines the neutrality of science not by arguing that scientific 'content is determined by its interests, but that it is limited by them' (p. 226). The content of science is value-laden by virtue of its commitment to particular epistemological values. Interest in power and control represents only one kind of value, made even more particular once one explicitly identifies the implied subjects and objects of such power. Finally, in noting that alternative values—alternative aims for science—have been championed not only by feminists but also by a number of other critics (including, for example, the theoretical chemist Ilya Prigogine), Tiles implicitly points to the possibility of new alliances for feminist philosophers of science.

Sandra Harding pursues the theme of multiplicity of values and continues the search for positive alternatives to prevailing conceptions of objectivity in her advocacy of what she calls 'strong objectivity'. Harding, long an advocate of standpoint theory, adds to that theory by opening it up to multiple marginalized subject positions (including race and ethnicity in her analysis), and by abandoning the notion of a single best-subject position. She also adds the requirement 'that the subject of knowledge be placed on the same critical, causal plane as the objects of knowledge' (p. 244). But she maintains from her earlier work the assertion that examination of the subjective assumptions that are carried into research can only be conducted by those who have hitherto been excluded from the position of epistemological subject.

The problem of objectivity is also the focus of Donna Haraway's essay. Dissatisfied both with the rejection of objectivity proposed by radical constructivism and with its revisions proposed by feminist standpoint and empiricist theories, she offers a productive reclamation of the visual metaphor. Where Keller and Grontkowski found in this metaphor historic support for the split between subject and object, Haraway recommends vision as a metaphor for healing that split. Keller and Grontkowski analyse the ways in which scientific theories of vision informed this metaphor; Haraway focuses on lessons to be drawn from new technologies and new biologies of vision. 'The "eyes" made available in modern technological sciences shatter any idea of passive vision; these prosthetic devices show us that all eyes, include our own organic ones, are active perceptual systems, building in translations and specific ways of seeing, that is, ways of life' (p. 254). So too, Haraway argues, 'Feminist objectivity is about limited location and situated knowledge, not about transcendence and splitting of the object. It allows us to become answerable to what we see' (p. 254).

Finally, Helen Longino notes the parallels between feminist arguments and recent trends in critical philosophy of science. Both converge on the conclusion that there is no pure or unconditioned subject position. Longino argues that such a recognition requires reconceiving knowledge (as distinct from opinion or belief) as social, that is, as the product of social interactions among members of a community and of interactions between them and the purported objects of knowledge rather than a matter of interactions only between an individual knower and the objects to be known. Revisioning objectivity, then, must involve not only reconceiving individuals' relationships to the world they seek to know, but articulating appropriate social structures and relations for the research contexts within which knowledge is pursued. She explores the further modifications in our concept of knowledge that such an approach demands, and argues, like Tiles, that scientific theories are evaluated by reference to the specific kinds of interactions with the natural world that they make possible.

All of these essays contribute crucially to current and future philosophical considerations about knowledge and objectivity. By raising important challenges to traditional conceptions of scientific knowledge, to characterizations of the knowing subject, the object of knowledge, and the relations between the two, they shift the terrain on which future philosophical debates about knowledge will have to be conducted. Though employing different analytic perspectives, they all help to show the interdependence of conceptions of subject and object, mind and body, epistemic and social values, how changes in one require changes in the others, and perhaps most importantly, demonstrate the possibility and feasibility of alternative approaches to the philosophical analysis of scientific knowledge. For virtually all our contributors, such alternatives must begin with the acknowledgement that scientific knowledge is immutably grounded, embodied, and partial; that its goals are subject to contestation; and that the dream of absolute, universal, and comprehensive truth is, like the dream of a final theory, just that i.e. a particular dream, the product of a particular historical and cultural moment.

This collection has been compiled in an effort to represent the theoretical dimensions of feminist science studies to date. This work has focused on identifying and understanding the operation of gender ideology directly in the biology of gender and through metaphor and analogy in other areas of the sciences. More ambitiously, it also seeks to reconceptualize knowledge and objectivity in ways that liberate both from particular gender ideologies. There is a great deal more to be done to carry this project forward. Among the most pressing issues we would identify two. One is the need for studies of language and gender in the physical sciences. Certainly gender will be found to operate differently in the physical than in the biological sciences. The ethnographic work of Sharon Traweek (1989) has demonstrated the intertwining of gender identities and community structures in one area of contemporary physics. Cohn's essay demonstrates the role of gender in linguistic and community structures among defence intellectuals. Detailed analyses of theories and practices in the physical sciences are needed to show the effects (if any) of such affective and social dimensions of the actual conduct of research and the accounts of the natural world these communities produce. The range of implications for work on gender and science will remain critically undetermined until such studies are done.

Secondly, it will hardly have escaped notice that the contributors to this volume are white and Western. While this reflects the history of the subject, the study of gender and science also remains incomplete without a deeper grasp of the interconnections between race, gender, and colonialist ideologies as manifested in the sciences. The essays by Stepan and by Tiles begin to show us where such a grasp might lead. The construction of a multinational and multicultural feminist academic community will, we hope, soon encompass science studies as well, bringing new questions, new perspectives, new transformations to this provocative and challenging subject.

These essays hardly constitute a final word on any problem in the philosophy of science; they do not even represent a final word on or comprehensive account of feminist critiques of science. We have had to make our choices, articulating what we take to be a significant trajectory of argument; and, as noted above, some choices were pre-made for us by the demographics of participation in this inquiry. To accommodate as broad a selection as permitted by our allotment of space, we have abridged all of the essays. We are grateful to the authors for their willingness to let us reprint their work in this form. Deletions are indicated by ellipses in the text. We are also grateful to Teresa Brennan and Susan James, the series editors, for their unqualified support of this project and to Tim Barton and Jenni Scott at Oxford University Press for their friendly and efficient handling of editorial dilemmas. Lauren Fox and Kerry Brooks helped us in the final stages of manuscript preparation.

References

Bleier, Ruth (1984), *Science and Gender* (Elmsford, NY: Pergamon).

Cohen, Estelle (1994), 'The Body as a Historical Category: Science and Imagination, 1660–1760' in Mary G. Winkler and Letha B. Cole (eds.) *The Good Body* (New Haven: Yale University Press).

Fausto-Sterling, Anne (1993), *Myths of Gender* (New York: Basic Books, 2nd edn.).

Hacking, Ian (1975), *Why Does Language Matter to Philosophy?* (Cambridge: Cambridge University Press).

Keller, Evelyn Fox (1995), *Refiguring Life: Metaphors of Twentieth-Century Biology* (New York: Columbia University Press).

Lloyd, Genevieve (1984), *The Man of Reason: 'Male' and 'Female' in Western Philosophy* (Minneapolis: University of Minnesota Press).

Russett, Cynthia (1989), *Sexual Science: The Victorian Construction of Womanhood* (Cambridge, Mass.: Harvard University Press).

Traweek, Sharon (1989), *Beamtimes and Lifetimes* (Cambridge, Mass.: Harvard University Press).

Part I. Early Statements

Women's Perspective as a Radical Critique of Sociology

Dorothy E. Smith

The women's movement has given us a sense of our right to have women's interests represented in sociology, rather than just receiving as authoritative the interests traditionally represented in a sociology put together by men. What can we make of this access to a social reality that was previously unavailable, was indeed repressed? What happens as we begin to relate to it in the terms of our discipline? We can of course think as many do merely of the addition of courses to the existing repertoire—courses on sex roles, on the women's movement, on women at work, on the social psychology of women and perhaps somewhat different versions of the sociology of the family. But thinking more boldly or perhaps just thinking the whole thing through a little further might bring us to ask first how a sociology might look if it began from the point of view of women's traditional place in it and what happens to a sociology which attempts to deal seriously with that. Following this line of thought, I have found, has consequences larger than they seem at first.

From the point of view of 'women's place' the values assigned to different aspects of the world are changed. Some come into prominence while other standard sociological enterprises diminish. We might take as a model the world as it appears from the point of view of the afternoon soap opera. This is defined by (though not restricted to) domestic events, interests, and activities. Men appear in this world as necessary and vital presences. It is not a woman's world in the sense of excluding men. But it is a women's world in the sense that it is the relevances of the women's place that govern. Men appear only in their domestic or private aspects or at points of intersection between public and private as doctors in hospitals, lawyers in their offices discussing wills and divorces. Their occupational and political world is barely present. They are posited here as complete persons, and they are but partial—as women appear in a sociology predicated on the universe occupied by men.

Reprinted from *Sociological Inquiry*, 44:1 (1974). By permission of the author and the University of Texas Press.

But it is not enough to supplement an established sociology by addressing ourselves to what has been left out, overlooked, or by making sociological issues of the relevances of the world of women. That merely extends the authority of the existing sociological procedures and makes of a women's sociology an addendum. We cannot rest at that because it does not account for the separation between the two worlds and it does not account for or analyse for us the relation between them. (Attempts to work on that in terms of biology operate within the existing structure as a fundamental assumption and are therefore straightforwardly ideological in character.)

The first difficulty is that how sociology is thought—its methods, conceptual schemes, and theories—has been based on and built up within, the male social universe (even when women have participated in its doing). It has taken for granted not just that scheme of relevances as an itemized inventory of issues or subject matters (industrial sociology, political sociology, social stratification, etc.) but the fundamental social and political structures under which these become relevant and are ordered. There is a difficulty first then of a disjunction between how women find and experience the world beginning (though not necessarily ending up) from their place and the concepts and theoretical schemes available to think about it in. Thus in a graduate seminar last year, we discussed on one occasion the possibility of a women's sociology and two graduate students told us that in their view and their experience of functioning in experimental group situations, theories of the emergence of leadership in small groups, etc. just did not apply to what was happening as they experienced it. They could not find the correlates of the theory in their experiences.

A second difficulty is that the two worlds and the two bases of knowledge and experience don't stand in an equal relation. The world as it is constituted by men stands in authority over that of women. It is that part of the world from which our kind of society is governed and from which what happens to us begins. The domestic world stands in a dependent relation to that other and its whole character is subordinate to it.

The two difficulties are related to one another in a special way. The effect of the second interacting with the first is to impose the concepts and terms in which the world of men is thought as the concepts and terms in which women must think their world. Hence in these terms women are alienated from their experience.

The profession of sociology is predicated on a universe which is occupied by men and it is itself still largely appropriated by men as their 'territory'. Sociology is part of the practice by which we are all governed and that practice establishes its relevances. Thus the institutions which lock sociology into the structures occupied by men are the same institutions which lock women into the situations in which they find themselves oppressed. To unlock the latter leads logically to an unlocking of the former. What follows then, or rather what then becomes possible—for it is of course by no means inevitable—is less a shift in the subject matter than a different conception of how it is or might become relevant as a means to understand our experience and the conditions of our experience (both women's and men's) in corporate capitalist society.

When I speak here of governing or ruling I mean something more general than the notion of government as political organization. I refer rather to that total complex of activities differentiated into many spheres, by which our kind of society is ruled, managed, administered. It includes that whole section which in the business world is called 'management'. It includes the professions. It includes of course government more conventionally defined and also the activities of those who are selecting, training, and indoctrinating those who will be its governors. The last includes those who provide and elaborate the procedures in which it is governed and develop methods for accounting for how it is done and predicting and analysing its characteristic consequences and sequences of events, namely the business schools, the sociologists, the economists, etc. These are the institutions through which we are ruled and through which we, and I emphasize this we, participate in ruling.

Sociology then I conceive as much more than ideology, much more than a gloss on the enterprise which justifies and rationalizes it and at the same time as much less than 'science'. The governing of our kind of society is done in concepts and symbols. The contribution of sociology to this is that of working up the conceptual procedures, models, and methods by which the immediate and concrete features of experience can be read into the conceptual mode in which the governing is done. What is actually observed or what is systematically recovered by the sociologist from the actualities of what people say and do, must be transposed into the abstract mode. Sociology thus participates in and contributes to the formation and facilitation of this mode of action and plays a distinctive part in the work of transposing the actualities of people's lives and experience into the conceptual currency in which it is and can be governed.

Thus the relevances of sociology are organized in terms of a perspective on the world which is a view from the top and which takes for granted the pragmatic procedures of governing as those which frame and identify its subject matter. Issues are formulated as issues which have become administratively relevant not as they are significant first in the experience of those who live them.

. . .

The governing processes of our society are organized as social entities constituted externally to those persons who participate in and perform them. The managers, the bureaucrats, the administrators, are employees, are people who are *used*. They do not own the enterprises or otherwise appropriate them. Sociologists study these entities under the heading of formal organization. They are put together as objective structures with goals, activities, obligations, etc., other than those which its employees can have as individuals. The academic professions are also set up in a mode which externalizes them as entities *vis-à-vis* their practitioners. The body of knowledge which its members accumulate is appropriated by the discipline as its body. The work of members aims at contributing to that body of knowledge.

. . .

An important set of procedures which serve to constitute the body of knowledge of the discipline as something which is separated from its practitioners are those known as 'objectivity'. The ethic of objectivity and the methods used in its practice are concerned primarily with the separation of the knower from what he knows and in particular with the separation of what is known from any interests, 'biases', etc., which he may have which are not the interests and concerns authorized by the discipline. I must emphasize that being interested in knowing something doesn't invalidate what is known. In the social sciences the pursuit of objectivity makes it possible for people to be paid to pursue a knowledge to which they are otherwise indifferent. What they feel and think about society can be taken apart from and kept out of what they are professionally or academically interested in.

The sociologist enters the conceptually ordered society when he goes to work. He enters it as a member and he enters it also as the mode in which he investigates it. He observes, analyses, explains, and examines as if there were no problem in how that world becomes observable to him. He moves among the doings of organizations, governmental processes, bureaucracies, etc., as a person who is at home in that medium. The nature of that world itself, how it is known to him and the conditions of its existence or his relation to it are not called into question. His methods of observation and inquiry extend into it as procedures which are essentially of the same order as those which bring about the phenomena with which he is concerned, or which he is concerned to bring under the jurisdiction of that order. His perspectives and interests may differ, but the substance is the same. He works with facts and information which have been worked up from actualities and appear in the form of documents which are themselves the product of organizational processes, whether his own or administered by him, or of some other agency. He fits that information back into a framework of entities and organizational processes which he takes for granted as known, without asking how it is that he knows them or what are the social processes by which the phenomena which correspond to or provide the empirical events, acts, decisions, etc., of that world, may be recognized. He passes beyond the particular and immediate setting in which he is always located in the body (the office he writes in, the libraries he consults, the streets he travels, the home he returns to) without any sense of having made a transition. He works in the same medium as he studies.

But like everyone else he also exists in the body in the place in which it is. This is also then the place of his sensory organization of immediate experience, the place where his coordinates of here and now before and after are organized around himself as centre; the place where he confronts people face to face in the physical mode in which he expresses himself to them and they to him as more and other than either can speak. It is in this place that things smell. The irrelevant birds fly away in front of the window. Here he has indigestion. It is a place he dies in. Into this space must come as actual material events, whether as the sounds of speech, the scratchings on the surface of paper which he constitutes as document, or directly anything he knows of the world. It has to happen here somehow if he is to experience it at all.

. . .

Women are outside and subservient to this structure. They have a very specific relation to it which anchors them into the local and particular phase of the bifurcated world. For both traditionally and as a matter of occupational practices in our society, the governing conceptual mode is appropriated by men and the world organized in the natural attitude, the home, is appropriated by (or assigned to) women (Smith 1973).

It is a condition of a man's being able to enter and become absorbed in the conceptual mode that he does not have to focus his

21

activities and interests upon his bodily existence. If he is to participate fully in the abstract mode of action, then he must be liberated also from having to attend to his needs, etc. in the concrete and particular. The organization of work and expectations in managerial and professional circles both constitutes and depends upon the alienation of man from his bodily and local existence. The structure of work and the structure of career take for granted that these matters are provided for in such a way that they will not interfere with his action and participation in that world. Providing for the liberation from the Aristotelian categories [time and space] of which Bierstedt (1966) speaks, is a woman who keeps house for him, bears and cares for his children, washes his clothes, looks after him when he is sick, and generally provides for the logistics of his bodily existence.

The place of women then in relation to this mode of action is that where the work is done to create conditions which facilitate his occupation of the conceptual mode of consciousness. The meeting of a man's physical needs, the organization of his daily life, even the consistency of expressive background, are made maximally congruent with his commitment. A similar relation exists for women who work in and around the professional and managerial scene. They do those things which give concrete form to the conceptual activities. They do the clerical work, the computer programming, the interviewing for the survey, the nursing, the secretarial work. At almost every point women mediate for men the relation between the conceptual mode of action and the actual concrete forms in which it is and must be realized, and the actual material conditions upon which it depends.

. . .

Women sociologists stand at the centre of a contradiction in the relation of our discipline to our experience of the world. Transcending that contradiction means setting up a different kind of relation than that which we discover in the routine practice of our worlds.

The theories, concepts and methods of our discipline claim to account for, or to be capable of accounting for and analysing the same world as that which we experience directly. But these theories, concepts, and methods have been organized around and built up out of a way of knowing the world which takes for granted the boundaries of an experience in the same medium in which it is constituted. It therefore takes for granted and subsumes without examining the conditions of its existence. It is not capable of analysing its conditions of our work. The social organization of the setting is not wholly available to us in its appearance. We bypass in the immediacy of the specific practical activity, a complex division of labour which is an essential pre-condition to it.

...

The matrix of direct experience as that from which sociology might begin discloses that beginning as an 'appearance' the determinations of which lie beyond it.

...

Women's situation in sociology discloses to her a typical bifurcate structure with the abstracted conceptual practices on the one hand and the concrete realizations, the maintenance routines, etc., on the other. Taking each for granted depends upon being fully situated in one or the other so that the other does not appear in contradiction to it. Women's direct experience places her a step back where we can recognize the uneasiness that comes in sociology from its claim to be about the world we live in and its failure to account for or even describe its actual features as we find them in living them. The aim of an alternative sociology would be to develop precisely that capacity from that beginning so that it might be a means to anyone of understanding how the world comes about for her and how it is organized so that it happens to her as it does in her experience.

Though such a sociology would not be exclusively for or done by women it does begin from the analysis and critique originating in their situation. Its elaboration therefore depends upon a grasp of that which is prior to and fuller than its formulation. It is a little like the problem of making a formal description of the grammar of a language. The linguist depends and always refers back to the competent speakers' sense of what is correct usage, what makes sense, etc. In her own language she depends to a large extent upon her own competence. Women are native speakers of this situation and in explicating it or its implications and realizing them conceptually, they have that relation to it of knowing it before it has been said.

The incomprehensibility of the determinations of our immediate local world is for women a particularly striking metaphor. It recovers an inner organization in common with their typical relation to the world. For women's activities and existence are determined outside them and beyond the world which is their 'place'. . . . As a sociologist then the grasp and exploration of her own experience as a method of discovering society restores to her a centre which in this enterprise at least is wholly hers.

own relation to its conditions because the sociologist as actual person in an actual concrete setting has been cancelled in the procedures which objectify and separate him from his knowledge. Thus the linkage which points back to its conditions is lacking.

For women those conditions are central as a direct practical matter, to be somehow solved in the decision to take up a sociological career. The relation between ourselves as practising sociologists and ourselves as working women is continually visible to us, a central feature of experience of the world, so that the bifurcation of consciousness becomes for us a daily chasm which is to be crossed, on the one side of which is this special conceptual activity of thought, research, teaching, administration and on the other the world of concrete practical activities in keeping things clean, managing somehow the house and household and the children, a world in which the particularities of persons in their full organic immediacy (cleaning up the vomit, changing the diapers, as well as feeding) are inescapable. Even if we don't have that as a direct contingency in our lives, we are aware of that as something that our becoming may be inserted into as a possible predicate.

It is also present for us to discover that the discipline is not one which we enter and occupy on the same terms as men enter and occupy it. We do not fully appropriate its authority, i.e. the right to author and authorize the acts and knowing and thinking which are the acts and knowing and thinking of the discipline as it is thought. We cannot therefore command the inner principles of our action. That remains lodged outside us. The frames of reference which order the terms upon which inquiry and discussion are conducted originate with men. The subjects of sociological sentences (if they have a subject) are male. The sociologist is 'he'. And even before we become conscious of our sex as the basis of an exclusion (*they* are not talking about *us*), we none the less do not fully enter ourselves as the subjects of its statements, since we must suspend our sex, and suspend our knowledge of who we are as well as who it is that in fact is speaking and of whom. Therefore we do not fully participate in the declarations and formulations of its mode of consciousness. The externalization of sociology as a profession which I have described above becomes for women a double estrangement.

There is then for women a basic organization of their experience which displays for them the structure of the bifurcated consciousness. At the same time it attenuates their commitment to a sociology which aims at an externalized body of knowledge based on an organization of experience which excludes theirs and excludes them except in a subordinate relation.

An alternative approach must somehow transcend this contradiction without re-entering Bierstedt's 'transcendental realm' (1966). Women's perspective, as I have analysed it here, discredits sociology's claim to constitute an objective knowledge independent of the sociologist's situation. Its conceptual procedures, methods, and relevances are seen to organize its subject matter from a determinate position in society. This critical disclosure becomes then the basis for an alternative way of thinking sociology.

. . .

We would reject, it seems to me, a sociology aimed primarily at itself. We would not be interested in contributing to a body of knowledge the uses of which are not ours and the knowers of whom are who knows whom, but generally male—particularly when it is not at all clear what it is that is constituted as knowledge in that relation. The professional sociologist's practice of thinking it as it is thought would have to be discarded. She would be constrained by the actualities of how it happens in her direct experience. Sociology would aim at offering to anyone a knowledge of the social organization and determinations of the properties and events of their directly experienced world. Its analyses would become part of our ordinary interpretations of the experienced world, just as our experience of the sun's sinking below the horizon is transformed by our knowledge that the world turns. (Yet from where we are it seems to sink and that must be accounted for.)

The only way of knowing a socially constructed world is knowing it from within. We can never stand outside it. A relation in which sociological phenomena are objectified and presented as external to and independent of the observer is itself a special social practice also known from within. The relation of observer and object of observation, of sociologist to 'subject', is a specialized social relationship. Even to be a stranger is to enter a world constituted from within as strange. The strangeness itself is the mode in which it is experienced.

. . .

To begin from direct experience and to return to it as a constraint or 'test' of the adequacy of a systematic knowledge is to begin from where we are located bodily. The actualities of our everyday world are already socially organized. Settings, equipment, 'environment', schedules, occasions, etc., as well as the enterprises and routines of actors are socially produced and concretely and symbolically organized prior to our practice. By beginning from her original and immediate knowledge of her world, sociology offers a way of making its socially organized properties first observable and then problematic.

Let me make it clear that when I speak of 'experience' I do not use the term as a synonym for 'perspective'. Nor in proposing a sociology grounded in the sociologist's actual experience, am I recommending the self-indulgence of inner exploration or any other enterprise with self as sole focus and object. Such subjectivist interpretations of 'experience' are themselves an aspect of that organization of consciousness which bifurcates it and transports us into mind country while stashing away the concrete conditions and practices upon which it depends. We can never escape the circles of our own heads if we accept that as our territory. Rather the sociologist's investigation of our directly experienced world as a problem is a mode of discovering or rediscovering the society from within. She begins from her own original but tacit knowledge and from within the acts by which she brings it into her grasp in making it observable and in understanding how it works. She aims not at a reiteration of what she already (tacitly) knows, but at an exploration through that of what passes beyond it and is deeply implicated in how it is.

. . .

If we address the problem of the conditions as well as the perceived forms and organization of immediate experience, we should include it in the events as they actually happen or the ordinary material world which we encounter as a matter of fact—the urban renewal project which uproots 400 families; how it is to live on welfare as an ordinary daily practice; cities as the actual physical structures in which we move; the organization of academic occasions such as that in which this paper originated. When we examine them, we find that there are many aspects of how these things come about of which we have little as sociologists to say. We have a sense that the events which enter our experience originate somewhere in a human intention, but we are unable to track back to find it and to find out how it got from there to here. Or take this room in which I work or that room in which you are reading and treat that as a problem. If we think about the conditions of our activity here, we could track back to how it is that there are chairs, table, walls, our clothing, our presence; how these places (yours and mine) are cleaned and maintained, etc. There are human activities, intentions, and relations which are not apparent as such in the actual material

References

Bierstedt, Robert (1966), 'Sociology and General Education', in Charles H. Page (ed.), *Sociology and Contemporary Education* (New York: Random House).

Gouldner, Alvin (1971), *The Coming Crisis in Western Sociology* (London: Heinemann Educational Books).

Smith, Dorothy E. (1973), 'Women, the Family and Corporate Capitalism', in M. L. Stephenson (ed.), *Women in Canada* (Toronto: Newpress).

2 Feminism and Science

Evelyn Fox Keller

In recent years, a new critique of science has begun to emerge from a number of feminist writings. The lens of feminist politics brings into focus certain masculinist distortions of the scientific enterprise, creating, for those of us who are scientists, a potential dilemma. Is there a conflict between our commitment to feminism and our commitment to science? As both a feminist and a scientist, I am more familiar than I might wish with the nervousness and defensiveness that such a potential conflict evokes. As scientists, we have very real difficulties in thinking about the kinds of issues that, as feminists, we have been raising. These difficulties may, however, ultimately be productive. My purpose in the present essay is to explore the implications of recent feminist criticism of science for the relationship between science and feminism. Do these criticisms imply conflict? If they do, how necessary is that conflict? I will argue that those elements of feminist criticism that seem to conflict most with at least conventional conceptions of science may, in fact, carry a liberating potential for science. It could therefore benefit scientists to attend closely to feminist criticism. I will suggest that we might even use feminist thought to illuminate and clarify part of the substructure of science (which may have been historically conditioned into distortion) in order to preserve the things that science has taught us, in order to be more objective. But first it is necessary to review the various criticisms that feminists have articulated.

The range of their critique is broad. Though they all claim that science embodies a strong androcentric bias, the meanings attached to this charge vary widely. It is convenient to represent the differences in meaning by a spectrum that parallels the political range characteristic of feminism as a whole. I label this spectrum from

Reprinted from *Signs: Journal of Women in Culture and Society*, 7/3 (1982). By permission of the University of Chicago Press.

right to left, beginning somewhere left of centre with what might be called the liberal position. From the liberal critique, charges of androcentricity emerge that are relatively easy to correct. The more radical critique calls for correspondingly more radical changes; it requires a re-examination of the underlying assumptions of scientific theory and method for the presence of male bias. The difference between these positions is, however, often obscured by a knee-jerk reaction that leads many scientists to regard all such criticism as a unit—as a challenge to the neutrality of science. One of the points I wish to emphasize here is that the range of meanings attributed to the claim of androcentric bias reflects very different levels of challenge, some of which even the most conservative scientists ought to be able to accept.

First, in what I have called the liberal critique, is the charge that is essentially one of unfair employment practices. It proceeds from the observation that almost all scientists are men. This criticism is liberal in the sense that it in no way conflicts either with traditional conceptions of science or with current liberal, egalitarian politics.

. . .

A slightly more radical criticism continues from this and argues that the predominance of men in the sciences has led to a bias in the choice and definition of problems with which scientists have concerned themselves. This argument is most frequently and most easily made in regard to the health sciences. It is claimed, for example, that contraception has not been given the scientific attention its human importance warrants and that, furthermore, the attention it has been given has been focused primarily on contraceptive techniques to be used by women.

. . .

This kind of criticism does not touch our conception of what science is, nor our confidence in the neutrality of science. It may be true that in some areas we have ignored certain problems, but our definition of science does not include the choice of problem—that, we can readily agree, has always been influenced by social forces. We remain, therefore, in the liberal domain.

Continuing to the left, we next find claims of bias in the actual design and interpretation of experiments. For example, it is pointed out that virtually all of the animal-learning research on rats has been performed with male rats.[1] Though a simple explanation is offered—namely, that female rats have a four-day cycle that complicates experiments—the criticism is hardly vitiated by the explanation. The implicit assumption is, of course, that the male

29

rat represents the species. There exist many other, often similar, examples in psychology.

. . .

Evidence for bias in the interpretation of observations and experiments is very easy to find in the more socially oriented sciences. The area of primatology is a familiar target. Over the past fifteen years women working in the field have undertaken an extensive re-examination of theoretical concepts, often using essentially the same methodological tools. These efforts have resulted in some radically different formulations. The range of difference frequently reflects the powerful influence of ordinary language in biasing our theoretical formulations. A great deal of very interesting work analysing such distortions has been done.[2]

. . .

These critiques, which maintain that a substantive effect on scientific theory results from the predominance of men in the field, are almost exclusively aimed at the 'softer', even the 'softest', sciences. Thus they can still be accommodated within the traditional framework by the simple argument that the critiques, if justified, merely reflect the fact that these subjects are not sufficiently scientific. Presumably, fair-minded (or scientifically minded) scientists can and should join forces with the feminists in attempting to identify the presence of bias—equally offensive, if for different reasons, to both scientists and feminists—in order to make these 'soft' sciences more rigorous.

It is much more difficult to deal with the truly radical critique that attempts to locate androcentric bias even in the 'hard' sciences, indeed in scientific ideology itself. This range of criticism takes us out of the liberal domain and requires us to question the very assumptions of objectivity and rationality that underlie the scientific enterprise. To challenge the truth and necessity of the conclusions of natural science on the grounds that they too reflect the judgement of men is to take the Galilean credo and turn it on its head. It is not true that 'the conclusions of natural science are true, and necessary, and the judgement of man has nothing to do with them',[3] it is the judgement of woman that they have nothing to do with.

The impetus behind this radical move is twofold. First, it is supported by the experience of feminist scholars in other fields of inquiry. Over and over, feminists have found it necessary, in seeking to reinstate women as agents and as subjects, to question the very canons of their fields. They have turned their attention, accordingly, to the operation of patriarchal bias on ever deeper levels of social structure, even of language and thought.

But the possibility of extending the feminist critique into the foundations of scientific thought is created by recent developments in the history and philosophy of science itself. As long as the course of scientific thought was judged to be exclusively determined by its own logical and empirical necessities, there could be no place for any signature, male or otherwise, in that system of knowledge. Furthermore, any suggestion of gender differences in our thinking about the world could argue only too readily for the further exclusion of women from science. But as the philosophical and historical inadequacies of the classical conception of science have become more evident, and as historians and sociologists have begun to identify the ways in which the development of scientific knowledge has been shaped by its particular social and political context, our understanding of science as a social process has grown. This understanding is a necessary prerequisite, both politically and intellectually, for a feminist theoretic in science.

Joining feminist thought to other social studies of science brings the promise of radically new insights, but it also adds to the existing intellectual danger a political threat. The intellectual danger resides in viewing science as pure social product; science then dissolves into ideology and objectivity loses all intrinsic meaning. In the resulting cultural relativism, any emancipatory function of modern science is negated, and the arbitration of truth recedes into the political domain. Against this background, the temptation arises for feminists to abandon their claim for representation in scientific culture and, in its place, to invite a return to a purely 'female' subjectivity, leaving rationality and objectivity in the male domain, dismissed as products of a purely male consciousness.

Many authors have addressed the problems raised by total relativism; here I wish merely to mention some of the special problems added by its feminist variant. They are several. In important respects, feminist relativism is just the kind of radical move that transforms the political spectrum into a circle. By rejecting objectivity as a masculine ideal, it simultaneously lends its voice to an enemy chorus and dooms women to residing outside of the realpolitik modern culture; it exacerbates the very problem it wishes to solve. It also nullifies the radical potential of feminist criticism for our understanding of science. As I see it, the task of a feminist theoretic in science is twofold: to distinguish that which is parochial from that which is universal in the scientific impulse, reclaiming for women what has historically been denied to them; and to legitimate those elements of scientific culture that have been denied precisely because they are defined as female.

It is important to recognize that the framework inviting what might be called the nihilist retreat is in fact provided by the very ideology of objectivity we wish to escape. This is the ideology that asserts an opposition between (male) objectivity and (female) subjectivity and denies the possibility of mediation between the two. A first step, therefore, in extending the feminist critique to the foundations of scientific thought is to reconceptualize objectivity as a dialectical process so as to allow for the possibility of distinguishing the objective effort from the objectivist illusion.

. . .

Rather than abandon the quintessentially human effort to understand the world in rational terms, we need to refine that effort. To do this, we need to add to the familiar methods of rational and empirical inquiry the additional process of critical self-reflection, . . . attending to the features of the scientific project that belie its claim to universality.

The ideological ingredients of particular concern to feminists are found where objectivity is linked with autonomy and masculinity, and in turn, the goals of science with power and domination. The linking of objectivity with social and political autonomy has been examined by many authors and shown to serve a variety of important political functions. The implications of joining objectivity with masculinity are less well understood. This conjunction also serves critical political functions. But an understanding of the sociopolitical meaning of the entire constellation requires an examination of the psychological processes through which these connections become internalized and perpetuated. Here psychoanalysis offers us an invaluable perspective, and it is to the exploitation of that perspective that much of my own work has been directed. In an earlier paper, I tried to show how psychoanalytic theories of development illuminate the structure and meaning of an interacting system of associations linking objectivity (a cognitive trait) with autonomy (an affective trait) and masculinity (a gender trait).[4] Here, after a brief summary of my earlier argument, I want to explore the relation of this system to power and domination.

Along with Nancy Chodorow and Dorothy Dinnerstein, I have found that branch of psychoanalytic theory known as object relations theory to be especially useful.[5] In seeking to account for personality development in terms of both innate drives and actual relations with other objects (i.e. subjects), it permits us to understand the ways in which our earliest experiences—experiences in large part determined by the socially structured relationships that form the context of our developmental processes—help to shape our conception of the world and our characteristic orientations to it. In particular, our first steps in the world are guided primarily by the parents of one sex—our mothers; this determines a maturational framework for our emotional, cognitive, and gender development, a framework later filled in by cultural expectations.

In brief, I argued the following: Our early maternal environment, coupled with the cultural definition of masculine (that which can never appear feminine) and of autonomy (that which can never be compromised by dependency) leads to the association of female with the pleasures and dangers of merging, and of male with the comfort and loneliness of separateness. The boy's internal anxiety about both self and gender is echoed by the more widespread cultural anxiety, thereby encouraging postures of autonomy and masculinity, which can, indeed may, be designed to defend against that anxiety and the longing that generates it. Finally, for all of us, our sense of reality is carved out of the same developmental matrix. As Piaget and others have emphasized, the capacity for cognitive distinctions between self and other (objectivity) evolves concurrently and interdependently with the development of psychic autonomy; our cognitive ideals thereby become subject to the same psychological influences as our emotional and gender ideals. Along with autonomy the very act of separating subject from object—objectivity itself—comes to be associated with masculinity. The combined psychological and cultural pressures lead all three ideals—affective, gender, and cognitive—to a mutually reinforcing process of exaggeration and rigidification. The net result is the entrenchment of an objectivist ideology and a correlative devaluation of (female) subjectivity.

This analysis leaves out many things. Above all it omits discussion of the psychological meanings of power and domination, and it is to those meanings I now wish to turn. Central to object relations theory is the recognition that the condition of psychic autonomy is double-edged: it offers a profound source of pleasure, and simultaneously of potential dread. The values of autonomy are consonant with the values of competence, of mastery. Indeed competence is itself a prior condition for autonomy and serves immeasurably to confirm one's sense of self. But need the development of competence and the sense of mastery lead to a state of alienated selfhood, of denied connectedness, of defensive separateness? To forms of autonomy that can be understood as protections against dread? Object relations theory makes us sensitive to autonomy's range of meanings; it simultaneously suggests the need to consider the corresponding meanings of competence. Under what circumstances does competence imply mastery of one's own fate and under what circumstances does it imply mastery over another's? In short, are control and domination essential ingredients of competence, and intrinsic to selfhood, or are they correlates of an alienated selfhood?

One way to answer these questions is to use the logic of the analysis summarized above to examine the shift from competence to power and control in the psychic economy of the young child. From that analysis, the impulse toward domination can be understood as a natural concomitant of defensive separateness—as Jessica Benjamin has written, 'A way of repudiating sameness, dependency and closeness with another person, while attempting to avoid the consequent feelings of aloneness.'[6] Perhaps no one has written more sensitively than psychoanalyst D. W. Winnicott of the rough waters the child must travel in negotiating the transition from symbiotic union to the recognition of self and other as autonomous entities. He alerts us to a danger that others have missed—a danger arising from the unconscious fantasy that the subject has actually destroyed the object in the process of becoming separate.

Indeed, he writes, 'It is the destruction of the object that places the object outside the area of control. . . . After "subject relates to object" comes "subject destroys object" (as it becomes external); then may come "*object survives* destruction by the subject". But there may or may not be survival.' When there is, 'because of the survival of the object, the subject may now have started to live a life in the world of objects, and so the subject stands to gain immeasurably; but the price has to be paid in acceptance of the ongoing destruction in unconscious fantasy relative to object-relating.'[7] Winnicott, of course, is not speaking of actual survival but of subjective confidence in the survival of the other. Survival in that sense requires that the child maintain relatedness; failure induces inevitable guilt and dread. The child is poised on a terrifying precipice. On one side lies the fear of having destroyed the object, on the other side, loss of self. The child may make an attempt to secure this precarious position by seeking to master the other. The cycles of destruction and survival are re-enacted while the other is kept safely at bay, and as Benjamin writes, 'the original self assertion is . . . converted from innocent mastery to mastery over and against the other'.[8] In psychodynamic terms, this particular resolution of preoedipal conflicts is a product of oedipal consolidation. The (male) child achieves his final security by identification with the father—an identification involving simultaneously a denial of the mother and a transformation of guilt and fear into aggression.

Aggression, of course, has many meanings, many sources, and many forms of expression. Here I mean to refer only to the form underlying the impulse toward domination. I invoke psychoanalytic theory to help illuminate the forms of expression that impulse finds in science as a whole, and its relation to objectification in particular. The same questions I asked about the child I can also ask about science. Under what circumstances is scientific knowledge sought for the pleasures of knowing, for the increased competence it grants us, for the increased mastery (real or imagined) over our own fate, and under what circumstances is it fair to say that science seeks actually to dominate nature? Is there a meaningful distinction to be made here?

In his work *The Domination of Nature* William Leiss observes, 'The necessary correlate of domination is the consciousness of subordination in those who must obey the will of another; thus properly speaking only other men can be the objects of domination'.[9] (Or women, we might add.) Leiss infers from this observation that it is not the domination of physical nature we should worry about but the use of our knowledge of physical nature as an instrument for the domination of human nature. He therefore sees the need for correctives, not in science but in its uses. This is his point of departure from other authors of the Frankfurt school, who assume the very logic of science to be the logic of domination. I agree with Leiss's basic observation but draw a somewhat different inference. I suggest that the impulse toward domination does find expression in the goals (and even in the theories and practice) of modern science, and argue that where it finds such expression the impulse needs to be acknowledged as projection. In short, I argue that not only in the denial of interaction between subject and other but also in the access of domination to the goals of scientific knowledge, one finds the intrusion of a self we begin to recognize as partaking in the cultural construct of masculinity.

The value of consciousness is that it enables us to make choices—both as individuals and as scientists. Control and domination are in fact intrinsic neither to selfhood (i.e. autonomy) nor to scientific knowledge. I want to suggest, rather, that the particular emphasis

35

Western science has placed on these functions of knowledge is twin to the objectivist ideal. Knowledge in general, and scientific knowledge in particular, serves two gods: power and transcendence. It aspires alternately to mastery over and union with nature. Sexuality serves the same two gods, aspiring to domination and ecstatic communion—in short, aggression and eros. And it is hardly a new insight to say that power, control, and domination are fuelled largely by aggression, while union satisfies a more purely erotic impulse.

To see the emphasis on power and control so prevalent in the rhetoric of Western science as projection of a specifically male consciousness requires no great leap of the imagination. Indeed, that perception has become a commonplace. Above all, it is invited by the rhetoric that conjoins the domination of nature with the insistent image of nature as female, nowhere more familiar than in the writings of Francis Bacon. For Bacon, knowledge and power are one, and the promise of science is expressed as 'leading to you Nature with all her children to bind her to your service and make her your slave',[10] by means that do not 'merely exert a gentle guidance over nature's course; they have the power to conquer and subdue her, to shake her to her foundations'.[11] In the context of the Baconian vision, Bruno Betelheim's conclusion appears inescapable: 'Only with phallic psychology did aggressive manipulation of nature become possible.'[12]

The view of science as an oedipal project is also familiar from the writings of Herbert Marcuse and Norman O. Brown.[13] But Brown's preoccupation, as well as Marcuse's, is with what Brown calls a 'morbid' science. Accordingly, for both authors the quest for a non-morbid science, an 'erotic' science, remains a romantic one. This is so because their picture of science is incomplete: it omits from consideration the crucial, albeit less visible, erotic components already present in the scientific tradition. Our own quest, if it is to be realistic rather than romantic, must be based on a richer understanding of the scientific tradition, in all its dimensions, and on an understanding of the ways in which this complex, dialectical tradition becomes transformed into a monolithic rhetoric. Neither the oedipal child nor modern science has in fact managed to rid itself of its preoedipal and fundamentally bisexual yearnings. It is with this recognition that the quest for a different science, a science undistorted by masculinist bias, must begin.

The presence of contrasting themes, of a dialectic between aggressive and erotic impulses, can be seen both within the work of individual scientists and, even more dramatically, in the juxtaposed writings of different scientists. Francis Bacon provides us with one model;[14] there are many others. For an especially striking contrast, consider a contemporary scientist who insists on the importance of 'letting the material speak to you', of allowing it to 'tell you what to do next'—one who chastises other scientists for attempting to 'impose an answer' on what they see. For this scientist, discovery is facilitated by becoming 'part of the system', rather than remaining outside; one must have a 'feeling for the organism'.[15] It is true that the author of these remarks is not only from a different epoch and a different field (Bacon himself was not actually a scientist by most standards), she is also a woman. It is also true that there are many reasons, some of which I have already suggested, for thinking that gender (itself constructed in an ideological context) actually does make a difference in scientific inquiry. Nevertheless, my point here is that neither science nor individuals are totally bound by ideology. In fact, it is not difficult to find similar sentiments expressed by male scientists. Consider, for example, the following remarks: 'I have often had cause to feel that my hands are cleverer than my head. That is a crude way of characterizing the dialectics of experimentation. When it is going well, it is like a quiet conversation with Nature.'[16] The difference between conceptions of science as 'dominating' and as 'conversing with' nature may not be a difference primarily between epochs, nor between the sexes. Rather, it can be seen as representing a dual theme played out in the work of all scientists, in all ages. But the two poles of this dialectic do not appear with equal weight in the history of science. What we therefore need to attend to is the evolutionary process that selects one theme as dominant.

Elsewhere I have argued for the importance of a different selection process.[17] In part, scientists are themselves selected by the emotional appeal of particular (stereotypic) images of science. Here I am arguing for the importance of selection within scientific thought—first of preferred methodologies and aims, and finally of preferred theories. The two processes are not unrelated. While stereotypes are not binding (i.e. they do not describe all or perhaps any individuals), and this fact creates the possibility for an ongoing contest within science, the first selection process undoubtedly influences the outcome of the second. That is, individuals drawn by a particular ideology will tend to select themes consistent with that ideology.

One example in which this process is played out on a theoretical level is in the fate of interactionist theories in the history of biology. Consider the contest that has raged throughout this century between organismic and particulate views of cellular organization—between what might be described as hierarchical and non-hierarchical theories. Whether the debate is over the primacy of the nucleus or the cell as a whole, the genome or the cytoplasm, the proponents of hierarchy have won out. One geneticist has described the conflict in explicitly political terms:

Two concepts of genetic mechanisms have persisted side by side throughout the growth of modern genetics, but the emphasis has been very strongly in favor of one of these. . . . The first of these we will designate as the 'Master Molecule' concept. . . . This is in essence the Theory of the Gene, interpreted to suggest a totalitarian government. . . . The second concept we will designate as the 'Steady State' concept. By this term . . . we envision a dynamic self-perpetuating organization of a variety of molecular species which owes its specific properties not to the characteristic of any one kind of molecule, but to the functional interrelationships of these molecular species.[18]

Soon after these remarks, the debate between 'master molecules' and dynamic interactionism was foreclosed by the synthesis provided by DNA and the 'central dogma'. With the success of the new molecular biology such 'steady state' (or egalitarian) theories lost interest for almost all geneticists. But today, the same conflict shows signs of re-emerging—in genetics, in theories of the immune system, and in theories of development.

I suggest that method and theory may constitute a natural continuum, despite Popperian claims to the contrary, and that the same processes of selection may bear equally and simultaneously on both the means and aims of science and the actual theoretical descriptions that emerge. I suggest this in part because of the recurrent and striking consonance that can be seen in the way scientists work, the relation they take to their object of study, and the theoretical orientation they favour. To pursue the example cited earlier, the same scientist who allowed herself to become 'part of the system', whose investigations were guided by a 'feeling for the organism', developed a paradigm that diverged as radically from the dominant paradigm of her field as did her methodological style.

In lieu of the linear hierarchy described by the central dogma of molecular biology, in which the DNA encodes and transmits all instructions for the unfolding of a living cell, her research yielded a view of the DNA in delicate interaction with the cellular environment—an organismic view. For more important than the genome as such (i.e. the DNA) is the 'overall organism'. As she sees it, the genome functions 'only in respect to the environment in which it is found'.[19] In this work the programme encoded by the DNA is itself subject to change. No longer is a master control to be found in a single component of the cell; rather, control resides in the complex interactions of the entire system. When first presented, the work underlying this vision was not understood, and it was poorly received.[20] Today much of that work is undergoing a renaissance, although it is important to say that her full vision remains too radical for most biologists to accept.

This example suggests that we need not rely on our imagination for a vision of what a different science—a science less restrained by the impulse to dominate—might be like. Rather, we need only look to the thematic pluralism in the history of our own science as it has evolved. Many other examples can be found, but we lack an adequate understanding of the full range of influences that lead to the acceptance or rejection not only of particular theories but of different theoretical orientations. What I am suggesting is that if certain theoretical interpretations have been selected against, it is precisely in this process of selection that ideology in general, and a masculinist ideology in particular, can be found to effect its influence. The task this implies for a radical feminist critique of science is, then, first a historical one, but finally a transformative one. In the historical effort, feminists can bring a whole new range of sensitivities, leading to an equally new consciousness of the potentialities lying latent in the scientific project.

Notes

1. I would like to thank Lila Braine for calling this point to my attention.
2. See, e.g. Donna Haraway, 'Animal Sociology and a Natural Economy of the Body Politic, Part I: A Political Physiology of Dominance'; and 'Animal Sociology and a Natural Economy of the Body Politic, Part II: The Past Is the Contested Zone: Human Nature and Theories of Production and Reproduction in Primate Behavior Studies', *Signs: Journal of Women in Culture and Society*, 4/1 (Autumn 1978), 21–60.
3. Galileo Galilei, *Dialogue on the Great World Systems*, trans. T. Salusbury, ed. G. de Santillana (Chicago: University of Chicago Press, 1953), 63.
4. Evelyn Fox Keller, 'Gender and Science', *Psychoanalysis and Contemporary Thought*, 1 (1978), 409–33.
5. Nancy Chodorow, *The Reproduction of Mothering: Psychoanalysis and the Sociology of Gender* (Berkeley: University of California Press, 1978); and Dorothy Dinnerstein, *The Mermaid and the Minotaur: Sexual Arrangements and Human Malaise* (New York: Harper & Row, 1976).

6. Jessica Benjamin has discussed this same issue in an excellent analysis of the place of domination in sexuality. See 'The Bonds of Love: Rational Violence and Erotic Domination', *Feminist Studies* 6/1 (Spring 1980), 144–74, esp. 150.
7. D. W. Winnicott, *Playing and Reality* (New York: Basic Books, 1971), 89–90.
8. Benjamin, 165.
9. William Leiss, *The Domination of Nature* (Boston: Beacon Press, 1974), 122.
10. B. Farrington, '*Temporis Partus Masculus:* An Untranslated Writing of Francis Bacon', *Centaurus*, 1 (1951), 193–205, esp. 197.
11. Francis Bacon, 'Description of the Intellectual Globe', in *The Philosophical Works of Francis Bacon,* ed. J. H. Robertson (London: Routledge & Sons, 1905), 506.
12. Quoted in Norman O. Brown, *Life against Death* (New York: Random House, 1959), 280.
13. Brown; and Herbert Marcuse, *One Dimensional Man* (Boston: Beacon Press, 1964).
14. For a discussion of the presence of the same dialectic in the writings of Francis Bacon, see Evelyn Fox Keller, 'Baconian Science: A Hermaphrodite Birth', *Philosophical Forum* 11/3 (Spring 1980), 299–308.
15. Barbara McClintock, private interviews, 1 Dec. 1978, and 13 Jan. 1979.
16. G. Wald, 'The Molecular Basis of Visual Excitation', *Les Prix Nobel en 1967* (Stockholm: Kungliga Boktryckerlet, 1968), 260.
17. Keller, 'Gender and Science'.
18. D. L. Nanney, 'The Role of the Cyctoplasm in Heredity', in William D. McElroy and Bentley Glass (eds.) *The Chemical Basis of Heredity* (Baltimore: Johns Hopkins University Press, 1957), 136.
19. McClintock, 1 Dec. 1978.
20. McClintock, 'Chromosome Organization and Genic Expression', *Cold Spring Harbor Symposium of Quantitative Biology*, 16 (1951), 13–44.

Reason, Science and the Domination of Matter

Genevieve Lloyd

INTRODUCTION

In a striking passage in *The Second Sex*, Simone de Beauvoir suggested that 'male activity', in prevailing over the 'confused forces of life', has subdued both Nature and woman.[1] The association between Nature and woman to which de Beauvoir here alludes has a long history in the self-definitions of Western culture. Nietzsche, with characteristic overstatement, suggested in a fragment on 'The Greek woman' that woman's closeness to Nature makes her play to the State the role that sleep plays for man.

In her nature lies the healing power which replaces that which has been used up, the beneficial rest in which everything immoderate confines itself, the eternal Same, by which the excessive and the surplus regulate themselves. In her the future generation dreams. Woman is more closely related to Nature than man and in all her essentials she remains ever herself. Culture is with her always something external, a something which does not touch the kernel that is eternally faithful to Nature.[2]

But in associating woman with sleep Nietzsche only pushed to its limits a long-standing antipathy between femaleness and active, 'male' Culture. The pursuit of rational knowledge has been a major strand in western culture's definitions of itself as opposed to Nature. It is for us in many ways equatable with Culture's transforming or transcending of Nature. Rational knowledge has been construed as a transcending, transformation or control of natural forces; and the feminine has been associated with what rational knowledge transcends, dominates, or simply leaves behind.

Reprinted with permission from *The Man of Reason* (London: Routledge, 1993), ch. 1.

FEMININITY AND GREEK THEORIES OF KNOWLEDGE

From the beginnings of philosophical thought, femaleness was symbolically associated with what Reason supposedly left behind— the dark powers of the earth goddesses, immersion in unknown forces associated with mysterious female powers. The early Greeks saw women's capacity to conceive as connecting them with the fertility of Nature. As Plato later expressed the thought, women 'imitate the earth'.[3] The transition from the fertility consciousness associated with cults of the earth goddesses to the rites of rational gods and goddesses was legendary in early Greek literature. It was dramatized, for example, in legends of the succession of cults at the site of the oracle at Delphi, incorporated into the prologue of Aeschylus' *Eumenides*, and elaborated as a story of conquest in Euripides' *Iphigenia in Tauris*. Euripides' version presented the transition as a triumph of the forces of Reason over the darkness of the earlier earth mysteries. The infant Apollo slays the Python which guards the old Earth oracle, thereby breaking the power of the Earth Goddess. She takes revenge by sending up dream oracles to cloud the minds of men with a 'dark dream truth'. But these voices of the night are stilled through the intervention of Zeus, leaving the forces of Reason installed at Delphi. Reason leaves behind the forces associated with female power.[4] What had to be shed in developing culturally prized rationality was, from the start, symbolically associated with femaleness.

These symbolic associations lingered in later refinements of the idea and the ideals of Reason; maleness remained associated with a clear, determinate mode of thought, femaleness with the vague and indeterminate. In the Pythagorean table of opposites, formulated in the sixth century BC, femaleness was explicitly linked with the unbounded—the vague, the indeterminate—as against the bounded—the precise and clearly determined. The Pythagoreans saw the world as a mixture of principles associated with determinate form, seen as good, and others associated with formlessness—the unlimited, irregular, or disorderly—which were seen as bad or inferior. There were ten such contrasts in the table: limit/unlimited, odd/even, one/many, right/left, male/female, rest/motion, straight/curved, light/dark, good/bad, square/ oblong. Thus 'male' and 'female', like the other contrasted terms, did not here function as straightforwardly descriptive classifications. 'Male', like the other terms on its side of the table, was construed as superior to its opposite; and the basis for this superiority was its association with the primary Pythagorean contrast between form and formlessness.

Associations between maleness and clear determination or definition persisted in articulations of the form—matter distinction in later Greek philosophical thought. Maleness was aligned with active, determinate form, femaleness with passive, indeterminate matter. The scene for these alignments was set by the traditional Greek understanding of sexual reproduction, which saw the father as providing the formative principle, the real causal force of generation, whilst the mother provided only the matter which received form or determination, and nourished what had been produced by the father. In the *Eumenides*, Aeschylus has Apollo exploit this contrast in the affirmation of father-right against mother-right in the moral assessment of Orestes' murder of his mother Clytemnestra, in vengeance of the murder of his father Agamemnon:

> The mother to the child that men calls hers
> Is no true life-begetter, but a nurse
> Of live seed. 'Tis the sower of the seed
> Alone begetteth. Woman comes at need,
> A stranger, to hold safe in trust and love
> That bud of her life—save when God above
> Wills that it die.[5]

Plato, in the *Timaeus*,[6] compared the role of limiting form to that of the father, and the role of indefinite matter to the mother; and Aristotle also compared the form–matter relation to that of male and female.[7] This comparison is not of any great significance for either of them in their explicit articulations of the nature of knowledge. But it meant that the very nature of knowledge was implicitly associated with the extrusion of what was symbolically associated with the feminine. To see the implications of this we must look in some detail at the way the form–matter distinction operated in Plato's theory of knowledge.

Knowledge, for Plato, involved a relation within human beings that replicates the relation in the rest of the world between knowable form and unknowable matter. Matching that separation on the side of the knower was a sharp distinction between mind—the principle which understands the rational—and matter, which has no part in knowledge. The knowing mind, like the forms which are its objects, transcends matter. Knowledge involved a correspondence between rational mind and equally rational forms. The idea that the world is itself suffused with Reason was present in much earlier Greek thought, but Plato greatly sophisticated it. In earlier thought, the intelligible object of knowledge was not sharply distinguished from the intelligence which knew it; the notion of *Logos* applied equally to both. Plato recast the idea of the world as mind-imbued in terms of the form–matter distinction; it was only in respect of form that the world was rational. The identification of rational thought and rational universe was not for him an unreflective assumption. It was achieved by deliberately downgrading matter to the realm of the non-rational, fortuitous and disorderly, while preserving for form the correspondence with rational, knowing mind.

. . .

Matter, with its overtones of femaleness, is seen as something to be transcended in the search for rational knowledge. It was the relation of master to slave, rather than that of man to woman, that provided the metaphors of dominance in terms of which the Greeks articulated their understanding of knowledge. But this Platonic theme recurs throughout the subsequent history of Western thought in ways that both exploit and reinforce the long-standing associations between maleness and form, femaleness and matter.

. . .

Plato's use of metaphors of dominance differed from later developments. In his theory, the dominance relation is seen as holding within the knower. The rightful dominance of mind over body, or of superior over inferior aspects of the soul, brings the knower into the required correspondence relations with the forms, which are in turn seen as superior to matter. On this model, knowledge is a contemplation of the eternal forms in abstraction from unknowable, non-rational matter. The symbolism of dominance and subordination occurs in the articulation of the process by which knowledge is gained. Knowledge itself is not seen as a domination of its objects, but as an enraptured contemplation of them.

Plato's picture has been highly influential in the formation of our contemporary ways of thinking about knowledge. But overlaid on it is a very different way of construing knowledge in terms of dominance—the model which receives its most explicit formulation, and has its most explicit associations with the male–female distinction, in the seventeenth century in the thought of Francis Bacon. On this model, knowledge itself is construed as a domination of Nature. This brings with it a different understanding of knowledge and its objects. To see the significance of the change it will be helpful first to look briefly at the transformation of Plato's version of the form–matter distinction in the thought of Aristotle.

Aristotle transformed Plato's form–matter distinction and its role in theory of knowledge; and with this transformation, the mind–body relationship also underwent a crucial change. In the *Metaphysics*, Aristotle hailed Plato's sophistication of the notion of form as a great advance over the primitive pre-Socratic cosmologies, which equated the basic principles of things with a single material element.[8] Plato's formal principles, Aristotle commented, were rightly set apart from the sensible. But he repudiated Plato's development of this insight into a dualism between a realm of change, apprehended through the senses, and a different realm of eternal forms. Aristotle brought the forms down from their transcendent realm to become the intelligible principles of changing, sensible things. The formal remains, for him, the proper object of necessary knowledge, and it is attained by the exercise of a purely intellectual faculty. But it is now grasped in the particular and sensible; it does not—as it did for Plato—escape into a distinct, supersensible realm. In Aristotle's own system, a dualism remained between what is sensed and what is grasped by Reason. But it no longer coincided with a distinction between changeable, created material things and uncreated, timeless, non-material forms. Aristotelian forms can function as intelligible principles of material things, and where they do, it is only in conjunction with matter that they can be regarded as existing at all. The mind–body relationship was accordingly transformed in the Aristotelian philosophy. The rational soul became the form of the body, and hence was no longer construed as the presence in human beings of a divine stuff which really belongs elsewhere. It was the intelligible principle of the body, not its prisoner; and rational knowledge was no longer construed as the soul's escape from the body.

. . .

The Aristotelian *rapprochement* of form and matter thus makes it possible for changeable material things to be proper objects of genuine knowledge. But within the Aristotelian framework this does not change the basic model of knowledge as a contemplation of forms. The form–matter distinction continues to operate, although it now holds within each object. Knowledge still involves the abstraction from matter of formal, intelligible principles, although these are no longer seen as located in a different realm from the sensible. The paradigm of knowledge is still the contemplation by a rational mind of something inherently mind-like, freed of matter.

Against the background of these contrasts, within a broader similarity, between the Platonic and Aristotelian views of knowledge, we can now see the significance of Bacon. In his thought, the gap between form and matter is completely closed. The split between knowable forms and unknowable matter is repudiated; and with it the model of knowledge as contemplation of forms. With this change, both the theme of dominance and the male–female distinction enter quite different relations with knowledge.

FRANCIS BACON: KNOWLEDGE AS THE SUBJUGATION OF NATURE

Bacon construed the mind's task in knowledge not as mere contemplation, but as control of Nature. This demanded a reconstruction of the proper objects of knowledge, bypassing the distinction between forms and matter. Forms—whether they be transcendent Platonic entities or Aristotelian abstract intelligible principles of material things—are appropriate objects of knowledge construed on a contemplative model. But if knowledge is construed as an instrument of control of Nature, its proper objects must be something more readily conceived as manipulable. Bacon's repudiation of the notion of form is thus closely connected with his conception of knowledge as power.

The only 'forms' Bacon was prepared to countenance as objects of scientific knowledge had very different relations to matter from those envisaged by either Plato or Aristotle. 'It is manifest,' he wrote in *The Advancement of Learning*, that Plato, 'a man of a sublime genius who took a view of everything as from a high rock, saw in his doctrine of ideas that "forms were the true object of knowledge".' But he 'lost the advantage of this just opinion by contemplating and grasping at forms totally abstracted from matter and not as determined in it'.[9] Bacon's closing of the gap between forms and matter went much further than that of Aristotle, amounting in effect to a total obliteration of the distinction. To understand physical nature, we must rather consider 'matter, its conformation, and the changes of that conformation, its own action, and the law of this action or motion; for forms are a mere fiction of the human mind, unless you will call the laws of action by that name'.[10] The understanding of physical Nature became for Bacon an understanding of the patterns in which matter is organized in accordance with mechanical laws.

The importance of this change goes beyond the progress of scientific knowledge.

In this new picture, the material world is seen as devoid of mind, although, as a product of a rational creator, it is orderly and intelligible. It conforms to laws that can be understood; but it does not, as the Greeks thought, contain mind within it. Nature is construed not by analogy with an organism, containing its intelligible principles of motion within it, but rather by analogy with a machine: as object of scientific knowledge, it is understood not in terms of intelligible principles enforming matter, but as mechanism. Bacon thus repudiated the model of knowledge as a correspondence between rational mind and intelligible forms, with its assumption that pure intellect could not distort reality. There are, he thinks, errors which 'cleave to the nature of the understanding'. 'For however men may amuse themselves, and admire, or almost adore the mind, it is certain, that like an irregular glass, it alters the rays of things, by its figure, and different intersections.'[11] The sceptics, rather than mistrusting the senses, should have mistrusted the 'errors and obstinacy of the mind', which refuses to obey the nature of things.[12] The mind itself should be seen as 'a magical glass, full of superstitions and apparitions'. The perceptions of the mind, no less than those of the senses, 'bear reference to man and not to the universe'.[13] Nature cannot be expected to conform to the ideas the mind finds within itself when it engages in pure intellectual contemplation. Knowledge must be painstakingly pursued by attending to Nature; and this attending cannot be construed in terms of contemplation.

Bacon, notoriously, used sexual metaphors to express his idea of scientific knowledge as control of a Nature in which form and matter are no longer separated. In Greek thought, femaleness was symbolically associated with the non-rational, the disorderly, the unknowable—with what must be set aside in the cultivation of knowledge. Bacon united matter and form—Nature as female and Nature as knowable. Knowable Nature is presented as female, and the task of science is the exercise of the right kind of male domination over her. 'Let us establish a chaste and lawful marriage between Mind and Nature,' he writes.[14] The right kind of nuptial dominance, he insists, is not a tyranny. Nature is 'only to be commanded by obeying her'.[15] But it does demand a degree of force: 'nature betrays her secrets more fully when in the grip and under the pressure of art than when in enjoyment of her natural liberty.'[16] The expected outcome of the new science is also expressed in sexual metaphors. Having established the right nuptial relationship, properly expressed in a 'just and legitimate familiarity betwixt the mind and things',[17] the new science can expect a fruitful issue from this furnishing of a 'nuptial couch for the mind and the universe'. From the union can be expected to spring 'assistance to man' and a 'race of discoveries, which will contribute to his wants and vanquish his miseries'.[18] The most striking of these sexual metaphors are in an early, strangely strident work entitled *The Masculine Birth of Time*. 'I am come in very truth', says the narrator in that work, 'leading to you Nature with all her children to bind her to your service and make her your slave.'[19]

My dear, dear boy, what I purpose is to unite you with *things themselves* in a chaste, holy and legal wedlock; and from this association you will secure an increase beyond all the hopes and prayers of ordinary marriages, to wit, a blessed race of Heroes or Supermen who will overcome the immeasurable helplessness and poverty of the human race, which cause it more destruction than all giants, monsters or tyrants, and will make you peaceful, happy, prosperous and secure.[20]

None of the elements of Bacon's account of knowledge is new. The idea that man has a rightful dominion—linked with his capacity for knowledge—over the rest of Nature goes back to the Genesis story, with reference to which Bacon named his *Great Instauration*. The proper direction of the arts and sciences, once the distortions of earlier false philosophies have been shed, is supposed to restore to man this rightful dominion lost through his original sin of pride. We have already seen the theme of mind's domination of matter in Plato's picture of knowledge as involving the subjection of the slave-like body and the soul. And the personification of Nature as female is no innovation. But Bacon brings all this together in a powerful new model of knowledge. The dominance relation—rather than holding between mind and body, or within the mind between different aspects of mental functioning—now holds between mind and Nature as the object of knowledge. Knowledge is itself the domination of Nature.

Today, Bacon's metaphors for the new science are inevitably seen in the light of contemporary preoccupations with the more negative aspects of science construed as human domination of Nature. But from his own perspective what is salient about the metaphors is quite different. They spring from a vision of the positive virtues of the new approach to science: the emphasis on sensory observation and experiment, the conviction that only an attentive observation of Nature, in conjunction with testing through experiments, will yield genuine knowledge. And they express the intellectual ideals which Bacon saw as implicit in this new science. These frequently disconcerting metaphors express two main points: first, that he who would know Nature must turn away from mere ideas and abstractions and painstakingly attend to natural phenomena; and, second, that this painstaking attention cannot be regarded as mere contemplation. Earlier philosophies, Bacon complains, being concerned with 'mere abstractions', only 'catch and grasp' at Nature, never 'seize or detain her'.[21] The Aristotelian philosophy has 'left Nature herself untouched and inviolate'; Aristotle 'dissipated his energies in comparing, contrasting and analysing popular notions about her'.[22] Bacon's demystified forms are always determined in matter, and understanding them is inseparable from the control and manipulation of Nature, although the practical and the speculative can, for convenience, be considered apart. This theme of the interconnections between knowledge and power is Bacon's main contribution to our ways of thinking about mind's relation to the rest of Nature. It is worth looking at it in more detail.

Philosophy, Bacon complains in the opening pages of *The Great Instauration*, has 'come down to us in the person of master and scholar, instead of inventor and improver'.[23] His own aim is not only the 'contemplative happiness', but the 'whole fortunes, and affairs, and powers, and works of men'. Man is both 'minister' and 'interpreter' of Nature; whence 'those twin intentions, human knowledge and human power, are really coincident'.[24] The speculative and the practical are distinguished from one another only as 'the search after causes' and 'the production of effects'; and these are in fact inseparable. The natural philosopher, as well as being a 'miner', digging out what lies concealed, also has the office of 'smelter'.[25] To understand forms is to be able to superinduce new natures on matter, which is the labour and aim of human power. Even that part of knowledge which may seem most remote from action—the understanding of forms—is ennobled by its role in releasing human power and leading it into an 'immense and open field of work'.[26]

The more we understand, the better our prospects of changing things; and these interconnections between knowledge and power are for Bacon so close as to amount to an identity. Truth and utility are equated, though not in a narrowly short-term utilitarian spirit. We should, Bacon urges, look for experiments that will 'afford light rather than profit',[27] confident in the expectation of long-term results from the better understanding of Nature. With that proviso understood, the 'practical' and the 'theoretical' are in fact the same:

49

'that which is most useful in practice is most correct in theory.'[28] The right analogy for the ideal science, then, is neither the activity of the ants, who merely heap up and use their store, nor that of the spiders, spinning out their webs; but rather that of the bee, who 'extracts matter from the flowers of the garden and the field, but works and fashions it by its own efforts'.[29] Practical results are not only the means to improve human well-being; they are the guarantee of truth.

The rule of religion, that a man should show his faith by his works, holds good in natural philosophy too. Science also must be known by works. It is by the witness of works, rather than by logic or even observation, that truth is revealed and established. Whence it follows that the improvement of man's mind and the improvement of his lot are one and the same thing.[30]

Seeing Bacon's equation of knowledge and power in this wider context highlights something which from our own historical perspective seems strange—that for Bacon the linking of knowledge with power was a return from the arrogance of earlier philosophies to intellectual humility. This is what his sexual metaphors are meant to convey. The control of Nature through science returns to man the rightful dominion which he lost through the sin of pride; and this dominion is regained precisely through the intellectual humility encapsulated in the new science. For Bacon, the endeavour to 'renew and enlarge the power and empire of mankind in general over the universe'[31] is a sound and noble ambition, involving chastity, restraint, and respect, not only on the part of Nature as chaste wife, but also on the part of her suitors. We have no right to expect Nature to come to us: 'Enough if, on our approaching her with due respect, she condescends to show herself.'[32] It is pride that has 'brought men to such a pitch of madness that they prefer to commune with their own spirits rather than with the spirit of nature'.[33] It is through pride, through wanting to be like gods, following the dictates of our own reason, that humanity has forfeited its rightful dominion over nature through true and solid arts.

Wherefore, if there be any humility towards the Creator, if there be any praise and reverence towards his works: if there be any charity towards men, and zeal to lessen human wants and sufferings; if there be any love of truth in natural things, any hatred of darkness, any desire to purify the understanding; men are to be entreated again and again that they should dismiss for a while, or at least put aside, those inconstant and preposterous philosophies which prefer these to hypotheses, have led experience captive, and triumphed over the works of God; that they should humbly and with a certain reverence draw near to the book of Creation; that they should there make a stay, that on it they should meditate, and that then washed and clean they should in chastity and integrity turn them from opinion. This is that speech and language which has gone out to all the ends of the earth, and has not suffered the confusion of Babel; this men must learn again, and, resuming their youth, they must become again as little children and deign to take its alphabet into their hands.[34]

But whatever may have been Bacon's conscious intent in describing scientific knowledge in terms of the male–female distinction, its upshot was to build a new version of the transcending of the feminine into the very articulation of the nature of science—this time with the emphasis on the malleability and tractability of matter. Matter is no longer seen as what has to be dominated in order to *attain* knowledge, but as the proper object of knowledge—now construed as the power to manipulate and transform. Malleability, rather than the eternal unchangeability of the forms, is the crucial feature of the objects of Baconian science. But this repudiation of the unknowability of matter did not shake the grip of earlier symbolic antitheses between femaleness and the activity of knowledge. On the contrary, it gave them a new and more powerful expression.

The transcending of the feminine was not an explicit feature of Greek theories of knowledge in their original form. But it was associated with knowledge through implicit associations of femaleness with matter, which pure intellect was supposed to transcend. Mind's domination of matter, as we have seen, was not explicitly associated with the male–female distinction, but rather with the master–slave relation. Early Greek associations of femaleness with matter did, none the less, influence the ways in which these theories of knowledge affected the philosophical imagination in later developments of the tradition. And in Bacon's metaphors the control of the feminine became explicitly associated with the very nature of knowledge.

How deep is the maleness of Bacon's expression of the nature of the new science? It may seem that it operates at a relatively superficial level. It is true that he unreflectively utilized associations between Nature and femaleness which abounded in his cultural tradition; and much of the content of his thought, as we have seen, can be explicated without the sexual metaphors. But the problem cannot be remedied by simply shedding superficial literary embellishments. The intellectual virtues involved in being a good Baconian scientist are articulated in terms of the right male attitude to the feminine: chastity, respect, and restraint. The good scientist is a gallant suitor. Nature is supposed to be treated with the respect appropriate to a femininity overlaid with long-standing associations with mystery—an awe, however, which is strictly contained. Nature is mysterious, aloof—but, for all that, eminently knowable and controllable. The metaphors do not merely express conceptual points about the relations between knowledge and its objects. They give a male content to what it is to be a good knower.

Both kinds of symbolism—the Greeks' unknowable matter, to be transcended in knowledge, and Bacon's mysterious, but controllable Nature—have played crucial roles in the constitution of the feminine in relation to our ideals of knowledge. The theme of the dominance of mind over body, or of intellect over inferior parts of the soul, was developed . . . in medieval versions of character ideals associated with maleness. And Bacon's connection of knowledge with power was developed in later ideas of Reason and progress.

Notes

1. de Beauvoir, S. (1949), *The Second Sex*, trans. H. M. Parshley, (Harmondsworth: Penguin Books, 1972), 97.
2. Nietzsche, F. (1871), 'The Greek woman', trans. M. A. Mügge, in O. Levy (ed.) (1911), *The Complete Works of Friedrich Nietzsche*, ii (London: T. N. Foulis), 22–3.
3. Plato, *Menexenus*, 238a.
4. For a discussion of the symbolic significance in Greek literature of the succession of cults at Delphi, see Harrison, J. (1912), *Themis: A Study of the Social Origins of Greek Religion* (Cambridge: Cambridge University Press), 385–96.
5. Aeschylus, *Eumenides*, 559, trans. G. Murray, in *The Complete Plays of Aeschylus* (London: Allen & Unwin, 1952), 235.
6. Plato, *Timaeus*, 50d.
7. Aristotle, *Metaphysics*, I, ch. 6, 998a 1–10, trans. R. McKeon, in *The Basic Works of Aristotle* (New York, Random House, 1941), 702.
8. Aristotle, op. cit. I, chs. 6–10, in McKeon, op. cit. 700–12.
9. Bacon, F. (1605), *The Advancement of Learning*, iii, ch. 4, in J. Devey (ed.) (1901), *The Physical and Metaphysical Works of Lord Bacon* (London: George Bell), 138.
10. Bacon, F. (1620a), *Novum Organum*, i, aphorism LI, in Devey, op. cit. 395.
11. Bacon, F. (1620b), *The Great Instauration*, 'Distribution of the Work', in Devey, op. cit. 15.
12. Bacon (1605), op. cit. v, ch. 2, in Devey, 188.
13. Ibid. v, ch. 4, in Devey, op. cit. 207.
14. Bacon, F., *The Refutation of Philosophies*, trans. B. Farrington, in *The Philosophy of Francis Bacon: An Essay on its Development from 1603 to 1609 with New Translations of Fundamental Texts* (Liverpool: Liverpool University Press, 1964), 131.
15. Bacon (1620a), op. cit., aphorism CXXIX, in Devey, op. cit. 447.
16. Bacon, F., *Thoughts and Conclusions*, in Farrington, op. cit. 99.

17. Bacon (1620*b*), op. cit., 'Announcement of the Author', in Devey, op. cit. 1.
18. Ibid., 'Distribution of the Work', in Devey, op. cit. 16.
19. Bacon, F. (1653), *The Masculine Birth of Time*, ch. 1, in Farrington, op. cit. 62.
20. Ibid., ch. 2, in Farrington, op. cit. 72.
21. Bacon (1620*a*), op. cit. i, aphorism CXXI, in Devey, op. cit. 441.
22. Bacon, *Thoughts and Conclusions*, sec. 13, in Farrington, op. cit. 83.
23. Bacon (1620*b*), op. cit., 'Preface', in Devey, op. cit. 3.
24. Ibid., 'Distribution of the Work', in Devey, op. cit. 20.
25. Bacon (1605), op. cit. iii, ch. 3, in Devey, op. cit. 122–3.
26. Ibid. iii, ch. 4, in Devey, op. cit. 140.
27. Bacon (1620*a*), op. cit. i, aphorism CXXI, in Devey, op. cit. 440.
28. Ibid. ii, aphorism IV, in Devey, op. cit. 451.
29. Ibid. i, aphorism XCV, in Devey, op. cit. 427.
30. Bacon, *Thoughts and Conclusions*, sec. 16, in Farrington, op. cit. 93.
31. Bacon (1620*a*), op. cit. i, aphorism CXXIX, in Devey, op. cit. 446.
32. Bacon, *The Refutation of Philosophies*, in Farrington, op. cit. 129.
33. Ibid., in Farrington, op. cit. 120.
34. Bacon, F. (1622–3) 'Preface' to *The History of the Winds*, in F. Bacon, *Works*, ii, collected and ed. J. Spedding, R. L. Ellis, and D. D. Heath, (London: Longman, 1858–74), 14–15; in Farrington, op. cit. 54–5.

Part II. Representations of Sex and Gender

then be the logical domination of a necessary but dangerous instinctual nature. Perhaps human beings found the key to control of sex, the source of and threat to all other kinds of order, in the categories of kinship. We learned that in naming our kind, we could control our kin. Only recently and tentatively have primatologists seriously challenged the indispensability of these sorts of explanations of nature and culture.

. . .

This paper is about the debate since approximately 1930 in primate studies and physical anthropology about human nature—in male bodies and female ones. The debate has been bounded by the rules of ordinary scientific discourse. This highly regulated space makes room for technical papers; grant applications; informal networks of students, teachers, and laboratories; official symposia to promote methods and interpretations; and finally, textbooks to socialize new scientists. The space considered in this paper does not provide room for outsiders and amateurs. One of the peculiar characteristics of science is thought to be that by knowing past regularities and processes we can predict events and thereby control them. That is, with our sciences—historical, disciplined forms of theorizing about our experience—we both understand and construct our place in the world and develop strategies for shaping the future.

How can feminism, a political position about love and power, have anything to do with science as I have described it? Feminism, I suggest, can draw from a basic insight of critical theory. The starting point of critical theory—as we have learned it from Marx, the Frankfurt school, and others—is that the social and economic means of human liberation are within our grasp. Nevertheless, we continue to live out relations of domination and scarcity. There is the possibility of overturning that order of things. The study of this contradiction may be applied to all of our knowledge, including natural science. The critical tradition insists that we analyse relations of dominance in consciousness as well as material interests, that we see domination as a derivative of theory, not of nature. A feminist history of science, which must be a collective achievement, could examine that part of biosocial science in which our alleged evolutionary biology is traced and supposedly inevitable patterns of order based on domination are legitimated. The examination should play seriously with the rich ambiguity and metaphorical possibilities of both technical and ordinary words. Feminists reappropriate science in order to discover and to define what is 'natural' for ourselves.[1] A human past and future would be placed in our

4 Animal Sociology and a Natural Economy of the Body Politic, Part II: The Past Is the Contested Zone

Donna Haraway

. . .

Primatology has focused on two major themes in interpreting the significance of animals for understanding human life—sex and economics, reproduction and production. The crucial transitions from natural to political economy and from biological social groups to the order of human kinship categories and systems of exchange have been basic concerns. These are old questions with complex relations to technical and ideological dimensions of biosocial science. Our understandings of both reproduction and production have double-edged possibilities. On one hand, we may reinforce our vision of the natural and cultural necessity of domination; on the other, we may learn to practise our sciences so as to show more clearly the now fragmentary possibilities of producing and reproducing our lives without overwhelming reliance on the theoretical categories and concrete practices of control and enmity.

Theories of animal and human society based on sex and reproduction have been powerful in legitimating beliefs in the natural necessity of aggression, competition, and hierarchy. In the 1920s, primate studies began to claim that all primates differ from other mammals in the nature of their reproductive physiology: Primates possess the menstrual cycle. That physiology was asserted to be fraught with consequences, often expressed in the fantasy-inspiring 'fact' of constant female 'receptivity'. Perhaps, many have thought and some have hoped, the key to the extraordinary sociability of the primate order rests on a sexual foundation of society, in a family rooted in the glands and the genes. Natural kinship was then seen to be transformed by the specifically human, language-mediated categories that gave rational order to nature in the birth of culture. Through classifying by naming, by creating kinds, culture would

Reprinted from *Signs: Journal of Women in Culture and Society*, 4/1 (1978). By permission of the University of Chicago Press.

hands. This avowedly interested approach to science promises to take seriously the rules of scientific discourse without worshipping the fetish of scientific objectivity.

My focus will be four sets of theories that emphasize the categories of reproduction and production in the tangled web of reconstruction of human nature and evolution. The first, centring on reproduction, was the work of Sir Solly Zuckerman. Born in 1904 in South Africa, he studied anatomy at the University of Cape Town, then earned his MD and BS at University College Hospital, London. He combined in complex and illuminating ways research interests in human paleontology and physical anthropology, reproductive physiology and the primate menstrual cycle, and broad zoological and taxonomic questions focused on primates. His social base included zoological gardens and research laboratories in British universities and medical schools; his training and career reflect intersections of the perspectives of anatomist, biochemist, anthropologist, clinician, administrator, and government science adviser.[2] He has been the architect of an extremely influential theory that sexual physiology is the foundation of primate social order. He also offered a variation of the theory of the origin of human culture in the hunting adaptation, which delineated crucial consequences for the division of labour by sex and the universal institution of the human family. In focusing on the sexual biology of monkeys, Zuckerman constructed a logic for setting the boundaries of human nature. In effect, Zuckerman claimed, the only universal for all the primates is the menstrual cycle. Therefore, only on that basis may we make valid comparisons of human and non-human ways of life.

A second set of theories stressing reproduction is that of Thelma Rowell, now at the University of California at Berkeley. She earned her doctorate in the early 1960s under Robert Hinde of Cambridge, the man who also supervised Jane Goodall's dissertation on chimpanzees. That period saw the beginning of a still continuing acceleration of publication based on long-term field observations of wild primates. Rowell's training was in zoology and ethology. Her first intention was to write her thesis on mammalian (hamster) communication, using the ethological approach worked out particularly by Niko Tinbergen at Oxford. Because Tinbergen then felt the methodology to be inappropriate to the non-stereotyped social communication of mammals, Rowell pursued her ideas at Cambridge under Hinde, a major synthesizer of American comparative psychology and Continental ethology. Rowell's research, which has used both traditions, has been concerned with primate communication, baboon menstrual cycle, comparison of naturalistic behaviour of monkeys with behaviour in captivity and in laboratory experimental situations, and mother–infant socialization systems.[3] An outspoken critic of the pervasive dominance concept, she has made her overriding theoretical concerns social role and stress.

Yet both Zuckerman and Rowell, who are very different, adopt varieties of biological and sociological functionalism that set limits on permitted explanations of the body and body politic. The most important is the functionalist requirement of an ultimate explanation in terms of equilibrium, stability, balance. Functionalism has been developed on a foundation of organismic metaphors, in which diverse physiological parts or subsystems are co-ordinated into a harmonious, hierarchical whole. Conflict is subordinated to a teleology of common interests. Both Zuckerman's and Rowell's explanations also reflect the ideological concerns of their society in complex ways, which can instruct feminist efforts to deal with biological and social theories.

[Editors' note: an extended discussion of Zuckerman and Rowell has been omitted for reasons of space.]
. . .

Stress as a global, multilayered concept embedded in functionalist explanation provides the critical tie between Thelma Rowell and Sherwood Washburn. The tie is represented by David Hamburg, now president of the Institute of Medicine of the National Academy of Sciences, formerly chairman of the Department of Sciences, formerly chairman of the Department of Psychiatry at the Stanford School of Medicine, and long-term collaborator with Sherwood Washburn in building primate studies around modern medical and evolutionary questions. In Hamburg's and Washburn's work, the darker side of functionalist explanation is starkly revealed; the metaphoric structure surrounding stress ceases to be more congenial than dominance. Hamburg has been the principal figure in evolutionary theories of emotional adaptive configuration, which lead to the notion of our obsolescent biology. Medical management of emotions maladaptive in 'modern society' seems justified to relieve pathological stress and maintain the social system. 'Modern society' itself seems given by some sort of technological imperative laid over our limiting biological heritage. Primate studies are motivated by, and in turn legitimate, the management needs of a stressed society. The animals model our limitations (adaptive breakdowns) and our innovations (tool use).

Social functionalism and evolutionary functionalism come together in the study of selection for behaviours and emotional patterns that maintain societies as successful breeding populations over time. The imperative is reproduction—of the social system and of the organisms who are its member-role actors. In general, animals have to like to do what they must do to survive in their evolutionary history. Evolutionary theory here joins a sociology of systems and a psychology of personality and emotion in modern versions of a pleasure calculus connected to the organic, motivational base of learning theory. Rowell summarizes: 'A zoologist, however, must always return to the question of selective advantages. . . . It is so very obvious that monkeys enjoy being together that we take it for granted. But pleasure like every other phenomenon of life is subject to, and the result of, evolutionary pressure—we enjoy a thing because our ancestors survived better and left more viable offspring than their relations who did not enjoy (and so seek) comparable stimuli. . . . This is speculation; but it is by research which examines the function of social systems of monkeys and other animals that we shall be able to understand fully their mechanisms. . . .'[4]

Washburn and Hamburg have shared the same analysis, but have applied it to another concept, again often perceived as a thing, in the vocabulary of meaning-laden scientific words: aggression, especially male aggression. Through this concept we must make a transition from explanations based on theories of reproduction to those based on production in human evolution and primate behaviour studies. Clearly, reproduction and production are complements, not opposites. But we must see how Washburn reached a 'man-the-hunter' theory from consideration of economic functions of the species, while Zuckerman traced primate order through reproductive physiology, and Rowell led us to understand the junction of sociological and evolutionary notions of reproducing systems.

In a 1968 paper, Washburn and Hamburg developed themes initiated in their collaboration in 1957, when Washburn spent a year as a fellow at the Stanford Center for Advanced Study in the Behavioral Sciences, and furthered in 1962–3 when Washburn and Hamburg organized at the Center a full year of conferences and collaboration among the exciting, worldwide community of primatologists. In 'Aggressive Behavior in Old World Monkeys and Apes',[5] the two collaborators introduced their work as part of the study of the forces that produced humankind. They wished to pay attention to unique human biology and unique conditions of human evolution. They saw aggression as a fundamental adaptation or functional complex, common to the entire primate order, including human beings. 'Order within most primate groups is maintained by a hierarchy, which depends ultimately primarily on the power of males. . . . Aggressive individuals are essential actors in the social system and competition between groups is necessary for species dispersal and control of local populations.'[6] The biology of aggression has been extensively studied and seems, they argue, to rest on similar hormonal and neural mechanisms, modified in primates, and especially in humans, by new brain complexes and extensive learning. In non-human primates, aggression is constantly rewarded, and, the authors maintain, aggressive individuals (males) leave more offspring. So they argue for selection of a system of coadapted genes involving complex feedback among motor anatomy, gestural anatomy, hormones, brain elements, and behaviour. Presumably, all parts of the aggressive complex evolve. The functions requiring aggression did not abate for humankind, Hamburg and Washburn believe. Protection, policing, and finally hunting all required a continued selection for male organisms who easily learned and enjoyed regulated fighting, torturing, and killing. 'Throughout most of human history societies have depended on young adult males to hunt, to fight, and to maintain the social order with violence. Even when the individual was co-operating, his social role could be executed only by extremely aggressive action that was learned in play, was socially approved, and was presumably gratifying.'[7]

But with the advance of civilization, this biology has become a problem. It is now often maladaptive because of our accelerating technological progress. Our bodies, with the old genetic transmission, have not kept pace with the new language-produced cultural transmission of technology. So now, when social control breaks down, we must expect to see pathological destruction. Hamburg and Washburn's examples here are Nazi Germany, the Congo, Algeria, and Viet Nam! The lesson is that we must face our *nature* in order to control it. 'There is a fundamental difficulty in the fact that contemporary human groups are led by primates whose evolutionary history dictates, through both biological and social transmission, a strong dominance orientation.'[8] This logic has been developed to posit a need for scientifically informed, rational controls to replace prescientific customs: 'But an aggressive species living by prescientific customs in a scientifically advanced world will pay a tremendous price in interindividual conflict and international war.'[9] The lesson here, the liberal scientist argues, is not to favour a particular social order—those are political and value questions—but to establish the preconditions for all advanced society, namely, scientific management of now inefficient, maladaptive, obsolescent biology. We are only one product, and one subject to considerable breakdown. On the personal level, psychiatric therapy is a species of repair work; on the social level, scientific policy dictates we use our skill to update our biology through social control. Our system of production has transcended us; we need quality control.

But before despairing that society is doomed to hierarchies and dominance relations regulated by scientific management, let us ask more closely what convinces Washburn and Hamburg that we, or at least males, have a woefully aggressive nature. After all, human males do not have the so-called fighting anatomy of many primate males—especially the dagger-like canines, associated threat gestures so appropriate for ethological analysis, great difference in male and female body size, and extra structures like a mane to enhance one's threatening aspect. Nor do we have appeasement gestures to placate aggressors. Why argue we do have an aggressive, authority-requiring brain? The line leading to genus *Homo*, Washburn judges, was bipedal and tool-using very early. Selection pressures favoured increased tool use, which in turn made possible the hunting way of life, evolution of a big brain, and language. Human males no longer fought with teeth and gestures but with words and handmade weapons. We lack big canines because we make knives and hurl insults. The selection pressures requiring aggression did not abate, but the structural basis for the function evolved in harmony with the whole adaptational complex of a new way of life. This argument itself relates to Washburn's basic reformulation of physical anthropology, beginning in the 1940s, as part of the synthetic theory of evolution, and to his successful efforts to promote primate behaviour studies in the study of human evolution.

Washburn earned his Ph.D. in physical anthropology at Harvard in 1940. His training was in traditional anthropometric methods and primate anatomy, and he taught medical anatomy at Columbia College of Physicians and Surgeons until 1947, when he moved to the University of Chicago. He had accompanied the 1937 Asia Primate Expedition, from which C. R. Carpenter produced the first monograph on gibbon behaviour and social system. But Washburn felt Carpenter then had little sense of the exciting possibilities of the concept of social system. His own task on the expedition was anatomical collecting, that is, shooting specimens. By the mid-1940s Washburn was practising physical anthropology as an experimental science; by 1950 he was developing a powerful programme for reinterpreting the basic concepts and methods of his field in harmony with the new population genetics, systematics, and palaeontology of Dobzhansky, Mayr, and Simpson. By 1958, he had a Ford Foundation grant to study the evolution of human behaviour in a complex manner, including provision for field studies of baboons in East Africa. A year later, now at Berkeley, he developed funding for one of the first experimental primate field-stations in the United States. From the beginning of his career, he lectured, wrote popular texts, made pedagogical films, reformed curricula on all educational levels, and promoted successful careers of now well-known figures in evolution and primatology.

This is not the place to explore the origins of Washburn's ideas, nor his organization of a very large research and education programme, but only to note essential features in relation to the hunting thesis and primate behaviour.[10] The purpose is to begin to recognize how Washburn's extraordinary career as a careful, experimental scientist has been part of the scientific and social controversies on human nature as the foundation for the human future. We must understand how Washburn could simultaneously be the coauthor of the article on evolution of aggression, an opponent of sociobiology, alternately a hero and villain for Robert Ardrey, a favourite of some Marxist feminists, and the teacher of both socio-biological feminist Adrienne Zihlman and of sociobiological sexist Irven De Vore. He is rightly all these things and unusually consistent and unified in his methods, theories, and practices. Perhaps the key to Washburn is that he has produced a fundamental theory with tremendous implications for the practice of many sciences and for the rules of speculative evolutionary reconstruction. In Kuhnian terms, Washburn seems to have something basic to do with scientific paradigms. In Marxist terms, he has to do with alienated theorizing of the established disorder.

Washburn's fundamental innovation in physical anthropology was evident in the publication of his widely reprinted papers, 'The New Physical Anthropology' and 'The Analysis of Primate Evolution with Particular Reference to Man'.[11] He applied the new population genetics to the study of primate evolution. For Washburn population genetics meant that the process of evolution was the crucial problem, not the fossil results. Therefore, selection and adaptation were his central concepts. Adaptive traits could only be interpreted by understanding conditions or forces capable of producing the traits. The first problem that confronted the physical anthropologist was how to identify a 'trait'. Washburn practised a new kind of theoretical and practical dissection of the body into 'functional complexes', whose meaning had to be sought in their action during life. For example, instead of measuring the nose, he analysed the forces in the central region of the face from chewing and growth. That task required model experimental systems of living animals. Instead of setting up scales of evolution based on brain enlargement, he analysed regions of the body involved in adaptive transformations related to locomotion, eating, and similar functions. In sum, 'the anatomy of life, of integrated functions, does not know the artificial boundaries which still govern the dissection of a corpse.'[12]

Washburn was part of a larger revolution in physical anthropology accompanied by discovery of new fossils, dating techniques, experimental possibilities, and more recently, molecular taxonomy. One of the revolution's central objects was the small-brained South African human-ape, *Australopithecus*. 'The discovery of the South African man-like apes, or small-brained men, has made it possible to outline the basic adaptation which is the foundation of the human radiation.'[13] The origin of the human radiation was like any other mammalian group's, but its consequences were decidedly novel. 'But the use of tools brings in a set of factors which progressively modifies the evolutionary picture. It is particularly the task of the anthropologist to assess the way the development of culture affected physical evolution.'[14] Evolutionary and social functionalism again come together, both, for Washburn, are analyses of meaning of living systems, of action, of ways of life. From the 1950s Washburn maintained that functional anatomy and the synthetic theory of evolution laid to rest forever the old conflicts of physical and social anthropology. He found Malinowski as central to his work as the geneticists. But as Zuckerman's Malinowski was the author of *Sex and Repression in Savage Society*, Washburn's was the writer of *A Scientific Theory of Culture and Other Essays*.[15]

In 1958, Washburn and his former student, Virginia Avis, contributed a paper to a symposium on Behavior and Evolution, which had been organized, beginning in 1953, to effect a synthesis of comparative psychology and the synthetic theory. Washburn's emphasis on the importance of behaviour made his interest in the psychological consequences of evolutionary adaptation natural. In that paper, 'The Evolution of Human Behavior,'[16] Washburn and Avis developed the consequences of the hunting adaptation, including enlarged curiosity and mobility, pleasure in the hunt and kill, and new ideas about our relation to other animals. Perhaps most important, 'hunting not only necessitated new activities and new kinds of co-operation but changed the role of the adult male in the group. . . . The very same actions which caused man to be feared by other animals led to more co-operation, food sharing, and economic interdependence within the group.'[17] The human way of life had begun.

From seeing behaviour first as motor activity and then as psychological orientations, it was a short, logical step to looking at behaviour as social system. Beginning in 1955, almost casually, Washburn investigated not only actions of individual organisms but of social systems. The baboon studies of Washburn and Irven De Vore, with all their emphasis on male roles in protection and policing as models of pre-adaptations to a human social system, were appropriate outgrowths of evolutionary functional anatomy. Differences between human and monkey society were always highlighted; Washburn never engaged in chain-of-being reconstructions. He looked at animal social systems the same way he looked at forces determining growth in kitten skulls—as model systems for particular problems in interpreting skull formation in fossils. His was an experimental, comparative biological science based on function. But the baboon model system drove home a lesson: troop structure came from dominance hierarchies of males. Hunting transformed such structures but only to produce the special roles of the co-operating male band. The reproductive function of females, and the social continuity of matrilines, remained a conservative pattern reinforced by bigger-headed, more dependent infants.

The classic paper which brings together anatomical, psychological, and social consequences of hunting in setting the rules for culture based on human nature is 'The Evolution of Hunting', by Washburn and C. S. Lancaster.[18] This paper has earned Washburn his poor reputation in socialist and feminist circles. Its appearance in a symposium emphasizing the hunting nature of man in the midst of years of challenge to sexual, economic, and political power is part of the social situation of contemporary evolutionary reconstruction. I am not claiming that Washburn is merely an ideologue; he is a good scientist and educator. That is the point. Interpreting human nature is a central scientific question for evolutionary functionalism. The past sets the rules for possible futures in the 'limited' sense of showing us a biology created in conditions supposedly favouring aggressive male roles, female dependence, and stable social systems appropriately analysed with functional concepts. Telling stories of the human past is a rule-governed activity. Washburn's science changed the rules of the game to require argument from conditions of production.

In 'Women in Evolution, Part I: Innovation and Selection in Human Origins',[19] Nancy Tanner and Adrienne Zihlman play by the new rules but tell of a different human nature, of different universals. They focus less on tools as such and more on the labour process, that is, on a new productive adaptation—gathering. They immediately place themselves within the recent population genetic developments of sociobiology. It explores a natural economy in terms of investment strategies for the increase of genetic capital. Yet Tanner and Zihlman deliberately appropriate sociobiology for feminist ends. They no more make themselves ideologues than Washburn has, but their practice of science is controversial for both internal reasons of debated evidence and argument and for political reasons. They do not, at any point, leave the traditional social space of science. They can stay there, in part, because sociobiology is not necessarily sexist in the sense that Irven De Vore or Robert Trivers have made it,[20] any more than the concept of stress necessarily leads to Hamburg's particular ideas on aggression and human obsolescence. Further, it is not easy to imagine what evolutionary theory would be like in any language other than classical capitalist political economy. No simple translation into other metaphors is possible or necessarily desirable. Tanner and Zihlman bring us face to face with fundamental questions that have barely been phrased, much less answered. How should we theorize our experience of the past and of 'nature' in new ways to build adequate concepts for scientific practice and social transformation? This question stands in a complicated relation with the internal craft rules for working within the natural sciences.

Tanner and Zihlman begin by announcing the goal of understanding human nature in terms of processes 'which shaped our physical, social, emotional, and cognitive characteristics'.[21] They note the obvious fact that the hunting thesis has largely ignored the behaviour and social activity of one of the two sexes, and is therefore deficient by ordinary criteria of evolutionary functionalism. Behaviour does not fossilize for either sex, so the problem is one of rational reconstruction, of choosing hypotheses.

Specifically, we hypothesize the development of gathering [both plant and animal material] as a dietary specialization of savanna living, promoted by natural selection of appropriate tool-using and bipedal behaviour. We suggest how this interrelates with the roles of maternal socialization in kin selection and of female choice in sexual selection. We emphasize the connections among savanna living, technology, diet, social organization, and selective processes to account for the transition from a primate ancestor to the emergent human species.[22]

This paper is clearly a normal outgrowth for Zihlman of her 1966 presentation on bipedal behaviour, in the context of hunting, to a Washburn-organized symposium of the American Anthropological Association. Titled 'Design for Man', the session included Hamburg on emotions as adaptational complexes and the problem of maladaptive, obsolete patterns.

Like Washburn, Tanner and Zihlman argue from animal model systems and from the most recent genetic theory applied to populations. They see chimpanzees as most closely similar of all living animals to the stem population that probably gave rise to apes and hominids. So chimpanzees make better mirrors, or models, than baboons for glimpses of evolution of the human way of life. The authors add to the traditional genetic parameters of the synthetic theory (drift, migration, and so on) the sociobiological genetic concepts of inclusive fitness, kin selection, sexual selection, and parental investment. The goal remains understanding changes in gene frequencies of populations from selection pressures operating on individuals. They note lots of tool use by chimpanzees, with a sex difference in the behaviour. Females make and use tools more often, although the males seem to hunt more readily. Rigid dominance hierarchies do not occur, although the concepts of high ranks and influence seem useful. The social structure is flexible, but not random. Social continuity seems to flow through continuing associations of females, their young, and associates.

The transitional population to hominids is imagined to have moved into the savanna, a new adaptive zone. 'A new way of life is initiated by a change in behaviour; the anatomical changes follow'.[23] The new behaviour was greatly enlarged dietary choice accompanied by tool use. Gathering was the early critical invention of hominids. Food sharing with ordinary social groups of females and offspring (including male sharing with these groups) resulted. Digging sticks, containers for food, and above all, carrying devices for babies were extremely likely early technological innovations related to the new diet and sharing habits. Knowledge of a wide range and seasons and habits of plants and animals became important. Selection pressure for symbolic communication increased. The predation dangers of the savanna were probably dealt with by cunning, not fighting, so hominids reduced needs for baboon-like dominance and male fighting anatomy. The flexible chimp social structure probably became even more opportunistic, allowing better understanding of the basis for human cultural diversity. Like Rowell, Tanner and Zihlman take every opportunity to emphasize human possibility and variety. Gathering of plants and animals was unlikely to maintain much selection pressure for an aggressive biology. Cognitive processes, on the other hand, were greatly elaborated in the new productive mode.

At this point, Tanner and Zihlman make use of mother-centred units to introduce kin and sexual selection and parental investment. New selection pressures put a premium on great sociability and co-operation. Babies were harder to raise, and bisexual co-operation would be useful. Males learned the friendly interaction patterns, even with strangers, which became crucial to the human way of life based on linguistic communities, small bands, and frequent out-breeding. But maintenance of a fighting anatomy including big canines and stereotyped threat gestures would be incompatible with the new functional behaviours. Females would mate more readily with friendly, non-threatening males. Female sexual choice has been shown to be general in mammalian groups, and the hominid stem was not likely to have been an exception. Two things leap at the reader who has followed Zuckerman's and Washburn's hunting arguments. First, female receptivity has been renamed female choice, with large genetic consequences. Second, the anatomy of the reduced canines is reinterpreted when different behaviour and different functions are postulated.

Tanner and Zihlman believe anthropology as a whole is better served by their different reconstruction, based on similar evidence.

Observers usually begin from their own perspective, and so inadvertently the question usually has been: how did the capacity and propensity for adult Western male behaviours evolve? This view point offers scant preparation for comprehending the wide range of variability in women's roles in non-Western societies or for analysing the changes in the roles of men and women which are currently occurring in the West.[24]

In other words, evolutionary reconstructions condition understanding of contemporary events and future possibilities. Tanner and Zihlman, in their interpretation of the tool-using adaptation, avoid telling a tale of obsolescence of the human body caught in a hunting past. The open future rests on a new past.

Focusing on the categories of reproduction and production, I have traced four major positions on human history and human nature. All were argued strictly within the boundaries of modern physiology, genetics, and social theory. All four hinged on the concept of function and recognized the 'liberal' doctrine of autonomy of nature and culture. It has been against the rules to argue from a position of biological reductionism. But the goal of each tale has been a picture of human universals, of human nature as the foundation for culture. Ironically, reconstructions of human nature useful to feminists were derived from two of the theories most despised by socialist-feminist thought: functionalism and sociobiology. They have been criticized as ideological justifications of unjust economic and political structures, as rationalizations for the reproduction of present relations of the body and body politic. Obviously, as Rowell, Tanner, and Zihlman show, these theories can be deployed for other ends: to stress human and animal variability, complexity, capacity for change. Feminists can engage seriously, then, in the biosocial debate from within the sciences.

However, we must be acutely aware of the dangers of taking the old rules to tell new tales. This is compatible with a larger refusal to pretend that science is either only discovery, which erects a fetish of objectivity, or only invention, which rests on crass idealism. We both learn about and create nature and ourselves. We must also see the biosocial sciences from the point of view of the process of resolving the contradiction between, or the gap between, human reality and human possibility in history. The purpose of the sciences of function is to produce both understanding of meaning and predictive means of control. They show both the given and the possible in a dialectic between the past and the future. Often, the future is given by the possibility of a past. Sciences also act as legitimating metalanguages that produce homologies between social and symbolic systems. That is acutely true for the sciences of the body and the body politic. In a strict sense, science is our myth. That claim does not in any way vitiate the discipline scientific practioners impose on each other to study the world. We can both know that our bodies, other animals, fossils, and what have you are proper objects for scientific investigation, and remember how historically determined is our part in the construction of the object. It is not an accident of nature that our social and evolutionary knowledge of animals, hominids, and ourselves has been developed in functionalist and capitalist economic terms. Feminists must not expect even arguments that answer clear sexist bias within the sciences to produce adequate final theories of production and reproduction as well. Such theories still elude us, because we are now engaged in a political-scientific struggle to formulate the rules through which we will articulate them. The terrain of primatology is the contested zone. The future is the issue.

Notes

1. I am indebted to Rayna Rapp, New School for Social Research, for clarification of this analysis of the meaning of the critical tradition.
2. For Zuckerman's own view of his career in science, see S. Zuckerman, *From Apes to Warlords: The Autobiography of Solly Zuckerman* (New York: Harper & Row, 1978), and *Beyond the Ivory Tower: The Frontiers of Public and Private Science* (New York: Taplinger, 1972).
3. See, for example, T. E. Rowell, 'Forest Living Baboons in Uganda', *Journal of Zoology*, 149 (1966), 344–64; 'Hierarchy in the Organization of a Captive Baboon Group', *Animal Behavior*, 14 (1966), 430–43; 'Baboon Menstrual Cycles Affected by Social Environment', *Journal of Reproduction and Fertility*, 21 (1970), 133–41.
4. Rowell, *Social Behavior of Monkeys*, 174, 180.
5. S. L. Washburn and David A. Hamburg, 'Aggressive Behavior in Old World Monkeys and Apes', in Phyllis Dolhinhow (ed.), *Primate Patterns* (New York: Holt, Rinehart & Winston, 1968).
6. Ibid. 282.
7. Ibid. 291.
8. Ibid. 295.
9. Ibid. 296.
10. Documentation to reconstruct his career, grants, students, and projects was courteously provided by Washburn from his files.
11. S. L. Washburn, 'The New Physical Anthropology', *Transactions of the New York Academy of Sciences*, ser. 2, 13/7 (1951), 298–304, and 'The Analysis of Primate Evolution with Particular Reference to Man', *Cold Spring Harbor Symposia on Quantitative Biology*, 15 (1951), 67–78.
12. Washburn, 'The New Physical Anthropology', 303.
13. S. L. Washburn, 'The Analysis of Primate Evolution', in William Howells (ed.), *Ideas on Human Evolution* (Cambridge, Mass.: Harvard University Press, 1962), 57.
14. Ibid. 158.
15. Bronislaw Malinowski, *A Scientific Theory of Culture and Other Essays* (Chapel Hill: University of North Carolina Press, 1944).
16. S. L. Washburn and V. Avis, 'The Evolution of Human Behavior', in Anne Roe and George Gaylord Simpson (eds.), *Behavior and Evolution*, (New Haven: Yale University Press, 1958), 421–36.
17. Ibid. 433–4.
18. S. L. Washburn and C. S. Lancaster, 'The Evolution of Hunting', in Richard Lee and Irven De Vore (eds.), *Man the Hunter* (Chicago: Aldine Publishing Co., 1968), 293–303.
19. Nancy Tanner and Adrienne Zihlman, 'Women in Evolution, Part I: Innovation and Selection in Human Origins', *Signs*, 1/3, pt. 1 (Spring 1976),

585–608. The second part, 'Subsistence and Social Organization in Early Hominids', appeared in *Signs*, 4/1 (Autumn 1978), 4-20.

20. R. I., Trivers, 'Parental Investment and Sexual Selection', in B. Campbell (ed.), *Sexual Selection and the Descent of Man* (Chicago: Aldine Publishing Co., 1972); R. B. Lee and I. De Vore (eds.), *Kalahari Hunter-Gatherers* (Cambridge, Mass.: Harvard University Press, 1976).

21. Tanner and Zihlman, 'Innovation and Selection', 585.

22. Ibid. 586.

23. Ibid.

24. Ibid. 608.

5 Body, Bias, and Behaviour: A Comparative Analysis of Reasoning in Two Areas of Biological Science

Helen E. Longino and Ruth Doell

INTRODUCTION

Our intention in this essay is to bring to light the variety in the ways masculine bias can express itself in the content and processes of scientific research. The discussion focuses on the two areas of evolutionary studies and endocrinological research into behavioural sex differences. Although both have attempted to construe the relation between sex and gender, the forms of these disciplines differ from one another in significant respects. Examining them together should lead to a broader, more subtle understanding of how allegedly extra-scientific considerations shape scientific inquiry.

While feminists have succeeded in alerting us to the existence of sexually prejudicial aspects of contemporary research that have implications for our understanding of sex differences, their critiques are dulled by a lack of adequate methodological analysis.[1] In her review of several collections of essays on sociobiology and hereditarianism, Donna Haraway remarks on the inconsistency of adopting a Kuhnian analysis of observation as theory- or paradigm-determined on the one hand, and asserting the incontrovertible existence of any fact on the other.[2] In the introductory and concluding sections of the feminist anthologies to which she refers,[3] we find the authors explaining sexist science both as bad science (it asks 'scientifically meaningless' questions and confuses correlation with causation) and as science as usual ('a product of the human imagination created from theory-laden facts', whose 'every theory is a self-fulfilling prophecy').[4] Now, if sexist science is bad science and reaches the conclusions it does because it uses poor methodology, this implies there is a good or better methodology that will steer us

Reprinted from *Signs: Journal of Women in Culture and Society*, 9/2 (1983). By permission of the University of Chicago Press.

away from biased conclusions. On the other hand, if sexist science is science as usual, then the best methodology in the world will not prevent us from attaining those conclusions unless we change paradigms. Only by developing a more comprehensive understanding of the operation of male bias in science, as distinct from its existence, can we move beyond these two perspectives in our search for remedies.

. . .

Since at least the time of Aristotle in the West, men have thought it important to justify their social dominance by appealing to ostensibly natural differences between males and females. As feminists we can identify the influence of patriarchal culture when we look at these theories and their proponents. Understanding *that* these theories incorporate a male bias and understanding *how* this can be so are, however, two distinct enterprises.

One way to approach the second problem is by examining the logical aspects of theory construction, particularly the determination of what counts as evidence and how such evidence is related to the hypotheses it is called on to support. These methodological categories can then be used as a probe to analyse the structure of inquiry in evolutionary and endocrinological studies. While we concentrate on the biology of evolution and behaviour, the analytic procedure we employ can be applied to other fields of inquiry as well.

A comprehensive understanding of bias would require, in addition, historical and sociological analysis of the institutions in which science is produced. Our analysis exposes the points of vulnerability in the logical structure of sciences to so-called external influences, such as culture, individual psychology, and institutional pressures. We shall argue that masculine bias expresses itself differently at different points in the chain of scientific reasoning (for instance in description of data and in inference from data), and that such differences require correspondingly different responses from feminists. Feminists do not have to choose between correcting bad science or rejecting the entire scientific enterprise. The structure of scientific knowledge and the operation of bias are much more complex than either of these responses suggests.

FACTS, EVIDENCE, AND HYPOTHESES

In our everyday world, we are surrounded by facts: singular facts (this ruby is red); general facts (all rubies are red); simple facts (the stove is hot); and complex facts (the hot stove burned my hand). Description of these facts is limited by the capacities of our sense organs and nervous systems as well as the contours of the language we use to express our perceptions. There is always much more going on around us than enters our awareness, not only because some of it occurs outside our sensory range or behind our backs, but also because in giving coherence to our experience we necessarily select certain facts and ignore others. The choice of facts to be explained by scientific means is a function of the reality constructed by this process of selection. What counts as fact—as reality—will thus vary according to culture, institutional perspective, and so on, making this process of selection one point of vulnerability to external influences.

Even the facts that enter our awareness are susceptible to a variety of descriptions. Accounts may be more or less concrete ('a rough-textured, grey, heavy cube' versus 'a building stone'); more or less value-laden ('she picked up the wallet' versus 'she stole the wallet'); and focused on different aspects ('grey' versus 'hard' versus 'cubical'). . . .

The possibility for multiple descriptions of a single reality means that, despite the ideals of scientific description, any given presentation of data may use terms that reflect social and cultural biases when other less value-laden or differently valued terms might do as well. This is another point of vulnerability to external factors.

An even smaller proportion of the change, flow, and movement in the world that enters our awareness functions as evidence. The category 'facts' and the category 'evidence' are not only not coextensive; they have their being in quite different ways. The structure of the facts we actually or potentially know is a function of our perceptual and intellectual structures. Evidence is constituted of facts taken in relation to something else—beliefs, hypotheses, theories. To speak of evidence is not to speak of bare facts or data awaiting an explanation. It is, instead, to confer on those facts an epistemic relevance to a belief, hypothesis, or theory.

. . .

Statements describing facts that are taken as evidence for hypotheses can be more or less direct consequences of the statements expressing those hypotheses. For example, the singular sentence, 'This swan is white', is a fairly direct consequence of the generalization, 'All swans are white.' In contrast, a statement describing discontinuities in the emission spectrum of hydrogen can be considered a consequence of a statement attributing different energy levels to the electron orbits of a hydrogen atom only in conjunction with a number of further assumptions that, for instance, assert a link between macroscopic phenomena like emission spectra and microscopic phenomena like atomic structure.[5] We will use the spatial term 'distance' to convey the logical notion of being more or less directly consequential. The less a description of fact is a direct consequence of the hypothesis for which it is taken to be evidence, the more distant that hypothesis is from its evidence. This distance that must be bridged between evidence and hypothesis provides yet another point of vulnerability to external influences.

Distinguishing between facts and evidence implies that which facts acquire scientific legitimacy will be a function of the theories under consideration. This in turn is determined by the explanatory needs of the scientific community, which are a function of specific social, institutional, and political goals. The concepts of evidence, and of the distance between a hypothesis and the evidence supporting it, are our primary analytical tools in the methodological examination of bias that follows. This approach facilitates comparisons within and between the areas investigated and helps make the operation of bias visible in scientific reasoning as well as in data collection and preparation.

THE ROLE OF EVIDENCE

Both evolutionary and endocrinological studies have as part of their purpose the elucidation of human nature. Evolutionary studies are concerned with the description of human descent: what happened—the temporal sequence of changes constituting the evolution of humans from an ancestral species—and how it happened—the mechanisms of evolution. Endocrinology attempts to articulate general laws that describe how hormones influence or control anatomical development, physiology, behaviour, and cognition. In the former case, researchers use the principles of the general synthetic theory of evolution to develop a historical reconstruction that can clarify what is human and what is natural about human nature. In the latter case, no history is sought; rather, universals about the natural, in the form of causal generalizations, are developed on the basis of contemporary observations, often made in experimental settings.

Both areas of inquiry take place within established research programmes, which address particular kinds of questions and abide by particular conventions as to how to go about answering those questions. We will discuss a parallel series of issues for both kinds of research: what questions are asked; what kinds of data are available, relevant, and appealed to as evidence for different types of questions; what hypotheses are offered as answers to those questions; what the distance between evidence and hypothesis is in each category; and finally, how these distances are traversed. Systematically assembling and analysing this material will make it possible to see some of the variety in the ways masculine bias functions in science.
. . .

In recent years, scientific reconstructions of human descent have centred around two focal images: man the hunter and woman the gatherer.[6] . . . Both accounts consider the development of tool use by early hominids a critical behavioural change. As an aid to survival it favoured the development of the bipedalism and upright posture necessary to wield tools effectively, and hence the anatomical changes that made new postures possible. . . . The androcentric account attributes the development of tool use itself to male hunting behaviour. By contrast, the gynecentric story explains the development of tool use as a function of female behaviour, viewing it as a response to the greater nutritional stress experienced by females during pregnancy, and later in the course of feeding their young through lactation and with foods gathered from the surrounding savannah. [Editors' note: the authors' extended discussion of evolutionary studies has been omitted.]
. . .

ENDOCRINOLOGICAL STUDIES OF SEX DIFFERENCES

The questions

Hormones regulate a variety of physiological functions. The role of sex hormones, the estrogens and androgens, in the development

77

and expression of sexually differentiated traits and functioning constitutes a small but intensively researched portion of the entirety of hormonal effects. Questions that have been studied regarding the relation of sex hormones to sexual differentiation can be grouped into three general categories corresponding to the three areas in which sexual differences are believed to be manifest: effects on anatomy and physiology, effects on temperament and behaviour, and effects on cognition. Within these areas are further distinctions concerning the timing and mechanism of hormonal activity which refer to whether a particular effect is due to fetal exposure that affects the organism's development, or to adolescent and adult exposure, which may have an activating or a permissive effect.[7]

The effects of androgens and estrogens on anatomical and physiological differentiation have been studied in relation to their role in the development of primary and secondary sex characteristics—reproductive organs, along with such traits as hair, voice, and body size—as well as their role in regulating postpubertal physiological functioning, including sperm production, cyclicity, and acyclicity. As research regarding the brain's relation to behaviour and physiology has become more sophisticated, the role of hormones in the development and organization of the brain has also become an object of inquiry. Studies of hormonal effects on behaviour have focused on sexual behaviour such as copulatory positioning and the frequency and timing of sexual activity, in addition to nonsexual but seemingly gender-linked behaviour and behavioural dispositions like fighting, aggression, dominance, submission, nurturance, grooming, and activity level in play. Questions about cognition address the possible influence of hormones in bringing about the well-known if not well-understood differences in verbal and spatial abilities between boys and girls. For reasons of space we shall limit our discussion to research in anatomy and physiology and in behaviour.

Data

Although there is a large amount of observational and experimental data available to serve as evidence for hypotheses regarding the relation of sex hormones to sexually differentiated characteristics, it is not highly consistent, nor is it all of the same quality. Information relevant to questions regarding anatomical and physical differentiation includes, first of all, observations of male and female body types and the correlation of these with higher and lower average levels of androgens and estrogens circulating in the body. Abnormalities in sex-linked anatomical and physiological characteristics have been correlated with deficiencies or excesses in hormonal levels, for example, the effects of castration on hair distribution and voice. In addition to data on humans, there are numerous animal studies determining the physiological effects of deliberate manipulation of hormone levels both perinatally and postnatally.

Animal experiments have also been performed to determine the effects of hormone levels on sexual behaviour, such as frequency of mounting, frequency of assuming the female mating posture, and increased or decreased female receptivity. One of the most extensively studied effects of hormonal activity on nonsexual behaviour involves the relation of testosterone levels to frequency of fighting behaviour in a variety of strictly controlled laboratory situations. In addition to the animal studies, there have been a number of attempts to correlate hormonal output with human behavioural differences. Commonly accepted stereotypes of sex-linked behaviours and their presumed correlation with different hormonal levels often provide the starting point and underlying context for more serious scientific explorations, despite the fact that the unrigorous and presumptive character of such stereotypes precludes their acceptance as genuine data. A more reliable source of information is found in controlled observations of the behaviour of individuals with hormonal irregularities. Among the groups studied are young women with CAH (congenital adrenocortical hyperplasia, a condition leading to the excess production of androgens during development, also referred to as AGS [adrenogenital syndrome]), young women exposed *in utero* to progestins, and male pseudohermaphrodites (genetic males with a female appearance until puberty, at which time they become virilized).

Hypotheses

Of the hypotheses articulated by researchers in this field, we shall restrict ourselves to only a few representative samples focused on sex differences in humans in the categories of inquiry we have distinguished.

The influence of sex hormones on the development of anatomically and physiologically sex-differentiated traits is generally acknowledged, and the mechanisms of development of the male and female reproductive systems are fairly well understood. Thus, it is widely accepted that during the third and fourth months of fetal life the bipotential fetus will develop the internal and then the external organs of the male reproductive system if exposed to androgen. Without such exposure, the fetus will develop female reproductive organs. The mechanisms of central nervous system development, while increasingly studied, are not yet as well understood. It has been hypothesized that androgen receptors play a primary role in sexual differentiation of the human brain, an assertion that rests on the assumption of sexually differentiated modes of brain organization.[8]

Hypotheses regarding the influence of sex hormones on behaviour trace their impact either to their perinatal organizing effects, or direct activating or permissive effects. In the arena of sexual behaviour, for example, several (largely unsuccessful) attempts have been made to attribute homosexuality to endocrine imbalances.[9] The area of non-sexual behaviour has also seen a proliferation of hypotheses. Steven Goldberg, along with other anthropologists, has argued that the social dominance of males is a function of hormonally determined behaviour.[10] Such theorists credit aggression with the capacity to determine one's position in hierarchical social structures and then attribute aggressive behaviour to the level of testosterone circulating in the organism. Even such thorough and non-patriarchal scholars as Eleanor Maccoby and Carol Jacklin endorse the claims that males on the whole exhibit higher levels of aggressive behaviour than females and that aggressive behaviour is a function of perinatal and circulating testosterone levels.[11] However, Maccoby and Jacklin are much more tentative about linking aggression with such phenomena as leadership and competitiveness.[12] Regarding other possible effects, Anke Ehrhardt has argued that gender-role or sex-dimorphic behaviour in humans, including 'physical energy expenditure', 'play rehearsal of parenting and adult behaviour', and 'social aggression', is influenced by perinatal exposure to sex hormones.[13]

Distance between evidence and hypotheses

As a model of reasoning in endocrinology, we can take the studies of hormonal influence on differentiation of the genitalia in humans. The current view that testosterone secreted by the fetal testis is required for normal male sex organ development and that female differentiation is independent of fetal gonadal hormone secretion is substantiated by observations in humans and by experimental data on a variety of other mammalian species. Among the human observations, most significant are those of persons affected by various hormonal abnormalities. Genetic males who lack intra-cellular androgen receptors and are thus unable to utilize testos-terone exhibit the female pattern of development of external genitalia. Genetic females exposed *in utero* to excess androgen, either as a result of progestin treatment of their mothers during pregnancy or due to their own adrenal abnormality, exhibit partial masculine development, including enlargement of the clitoris and incomplete fusion of the labia. These observations support the hypothesis that no particular hormonal secretion from the fetal gonad is required for female development, whereas exposure of the primordial tissues to testosterone or one of its metabolites at the appropriate time is both necessary and sufficient for masculine development of the sex organs. This inference is further corrobo-rated by experimental data in a variety of mammalian species whose reproductive anatomy and physiology are analogous to those of humans. For instance, castration of male fetuses *in utero* invari-ably results in their developing a female appearance.

In contrast to the security of the hypothesis regarding sex organ development are issues regarding the biochemical pathways testos-terone follows in producing its physiological effects. Because the exact mechanism of hormonal action at the cellular level is only partially understood, it is not yet certain how testosterone or one of its metabolites acts in the cell nucleus.

. . .

The relation between data and hypotheses becomes much more complex in attempts to link hormonal levels with behaviour. The inferential steps in Ehrhardt's work on young women with CAH provide an interesting illustration of this complexity. Unlike some of the authors exploring this topic, Ehrhardt is directly engaged in aspects of the research that forms the basis of her thinking. In addi-tion, since she is concerned with the relation between prenatal hor-mone exposure and later behaviour, there is no question of hormone levels being an effect of behaviour rather than vice versa. From the point of view of hereditarian theories of gender, Ehrhardt's work, if sound, would indicate a mechanism that medi-ates between the genotype and its behavioural expression. All these factors confer on her work a pivotal significance.[14]

The data Ehrhardt brings to her line of reasoning include both observations of humans and experimentation with rats. The human observations follow girls affected by CAH, using their

81

female siblings as controls. She documents the fact of the girls' prenatal exposure to greater than normal quantities of androgens and
evaluates observations of their behaviour as children and adoles-
cents. It is important to remember that these girls were born with
genitalia that were surgically altered in later life and that they
require lifelong cortisone treatment. The majority were said to
exhibit 'tomboyism', operationally characterized as preference for
active outdoor play, preference for male over female playmates,
greater interest in a career than in housewifery, as well as less interest in small infants and less play rehearsal of motherhood roles than
that exhibited by unaffected females. One problem with these
behavioural observations concerns Ehrhardt's method of data collection. Because these observations were obtained from the girls
themselves and from parents and teachers who knew of the girls'
abnormal physiological condition, it is difficult to know how much
the reports are influenced by observers' expectations.

Leaving this problem aside, let us proceed with the reconstruction of Ehrhardt's line of reasoning. She advances the hypothesis
that human gender-role behaviour, that is, behaviour considered
appropriate to one gender or the other but not to both, is
influenced by prenatal exposure to sex hormones. This hypothesis
is a generalization of the specific explanation offered for CAH
women, namely, that engaging in a degree of gender-role behaviour
thought inappropriate to their chromosomal and anatomical sex is
a function of their prenatal exposure to excessive amounts of
androgen. It is significant that Ehrhardt's treatment of the CAH
women does not consider the observational data (available in some
quantity) indicating the effect of early environmental factors in
shaping alleged gender-role behaviours.

What justifies the attribution of the CAH girls' behaviour to
physiological rather than environmental factors? To begin to
answer this question, Ehrhardt appeals to research on other mammalian species that seems to show the hormonal determination of
certain behaviours. The premise that physiological and behavioural
phenomena are continuous throughout mammalian species allows
her to assign the allegedly sex-inappropriate activities of CAH
women the status of evidence for the hormonal determination of
behaviour. When she cites recent research on rodent brains and
behaviour to support her interpretation of the human studies, she
is assuming that the rodent and human situations are similar
enough that demonstration of a causal mechanism in one species is
adequate to support the inference from correlation to causation in

the other. There are several recognized difficulties with this assumption. Obviously the human brain is much more complex than the rodent brain. Second, experiments with rodents all involve single factor analysis, while human situations, including that of the CAH girls, are always interactive. Finally, some of the rodent experiments are equivocal in their support of the hormonal determination of rodent behaviour.

In addition to these problems in extrapolating from the results of non-human animal experiments to humans, alternative explanations are not ruled out by the data Ehrhardt presents. More sociologically and culturally oriented studies have depicted the kind of behaviour exhibited by the CAH girls as an outcome of social and environmental factors.[15] Such studies supply a framework within which the girls' behaviour can be seen as evidence for certain early environmental influences. Equally plausible is the hypothesis that the girls' behaviour is a deliberate response to their situation as they perceive it. Because the alleged tomboyism of CAH girls is not unique to them but shared by many young women without demonstrated hormonal irregularities, the difference between the CAH girls and the control group is as likely a function of environment or self-determination as a direct product of their hormonal states. Support for the hormonal explanation in the CAH case must include arguments ruling out such alternative explanations. To date such arguments have not been provided.[16]

The considerable distance between evidence and hypotheses regarding the hormonal determination of behavioural sex differences contrasts sharply with the close fit between the two in the case of anatomical sexual differentiation. The human reproductive system is not significantly different from that of non-human mammals and the mechanism of anatomical differentiation in the latter is clearly known. While the course of development in the human embryo has not been observed directly as it has been in other species, the hypothesis of hormonal determination in humans can be seen as a causal generalization from instances where hormonal and anatomical abnormalities are correlated in humans in accordance with Mill's rules of agreement and difference in causal reasoning.[17] Because human behavioural dispositions cannot be exclusively associated with prenatal hormonal levels and receptors in the same way as anatomical conditions, the argument regarding gender-dimorphic behaviour fails Mill's agreement test and falls back on animal modelling to give the data relevance as evidence in order to bridge the gap between data and hypotheses. Animal modelling, we have argued, precludes any generalization from the situation of the CAH women because it fails to support the specific inference of causation in that case. This leaves the choice of a physiological or an environmental explanation for behaviour (or some alternative to the nature/nurture dichotomy), like the choice of framework in evolutionary studies, subject to the pre-conceived ideas and values of the researcher.

UNDERSTANDING MALE BIAS

We consider in this section the implications of our analysis for understanding the expression of male bias in the development of theory. While there are obvious interconnections between the types of research we have discussed, there are also some significant discontinuities and differences.

We noted at the outset that the aims of evolutionary and endocrinological studies are quite distinct. Evolutionary studies attempt to reconstruct prehistory by recovering particulars and relating them in order to describe the development of a particular species, *Homo sapiens.* On the basis, generalizations concerning the interrelation of various aspects of human existence become possible, but their production is not an immediate objective. In contrast, the goal of neuroendocrinological research is to discover the hormonal substrates of certain behaviours by developing causal or quasi-causal generalizations relating the two. To the extent that evolutionary studies are believed to reveal certain behaviours or behavioural dispositions as expressions of human nature and neuroendocrinological studies to reveal hormonal determinants of those behaviours, the otherwise quite disparate aims of these fields intersect.

At a certain historical phase in both lines of inquiry, we find researchers attempting to achieve precisely this kind of synthesis. Evolutionary studies undertaken within a certain framework have been held to demonstrate that the sexual division of labour observable in some contemporary human societies has deep roots in the evolution of the species. Some contend that man-the-hunter stories of males going off together to hunt large animals while females stayed home to nurture their young prefigure contemporary Western middle-class social life in which men engage in public and women in domestic affairs.[18] If these broadly described behaviours or behavioural tendencies could be correlated with the more particularized behaviours and behavioural dispositions studied by neuroendocrinology, a picture of biologically determined human universals would emerge. Evolutionary studies would provide the universals—gender and sex roles that have remained fundamentally constant throughout the history of the species—while neuroendocrinology provided the biological determination—the dependence of these particular behaviours or behavioural dispositions on prenatal hormone distribution. We have employed a logical analysis focused on the character and role of evidence in these areas of inquiry to show that neither claim need be accepted. Their conjunction obviously can fare no better.

It is instructive to note not only the ways these inquiries intersect but also their distinguishing features, particularly in their expression of masculine bias.[19] In evolutionary studies assigning key significance to man the hunter, androcentric bias is expressed directly in the framework within which data are interpreted: chipped stones are taken as unequivocal evidence of male hunting only in a framework that sees male behaviour as central not only to the evolution of the species but to the survival of any group of its members. In current neuroendocrinological studies, because there is no comparably explicit androcentric framework for the interpretation of data, the choice of a physiological framework is not directly related to androcentric bias. Feminists, however, have identified sexist bias in the endocrinologists' search for physiological rather than environmental explanations. One reason for attributing masculine bias to this preference is the potential, noted above, for linking physiological explanations with the androcentric evolutionary account to produce a picture of a biologically determined human nature. This possibility has raised concern that some will see in a biologically determined human nature which includes behavioural sex differences sufficient justification for maintaining social and legal inequalities between the sexes. On a personal level, many fear that men (individually or *en masse*) will use such a view to buttress their resistance to change. Yet these political interests are served only if one assumes that allegedly masculine characteristics are preferable or superior to allegedly feminine characteristics, that the allegedly physiological basis of these attributes makes them immutable, and that such differences provide adequate grounds for female subordination. The popularity of these assumptions does not mean they can withstand critical scrutiny. Nevertheless, their prevalence explains why feminists are alarmed by attempts to provide physiological explanations for behaviour.

Certainly some proponents of the physiological view are influenced by sexist motivations, either their own or those of the research directors, review committees, journal editors, and referees who create the climate in which research is produced and received. But is the physiological project itself sexist? With respect to the methodological categories of analysis connected with evidence, we can look at both the description of data presented as evidence and the assumptions mediating inferences from data to hypotheses.

Physiological explanations are clearly sexist in their description of assumed gender-dimorphic behaviour. Using a term like 'tomboyism' to describe the behaviour of CAH girls reflects an initial acceptance of social prescriptions for sex-appropriate behaviour.[20] This body of research is also androcentric and ethnocentric in its assumption that behavioural differences apparent in the investigator's culture represent human universals. However, these are problems of description and presentation; choosing a less value-laden term than 'tomboy' might allow for the description of genuine differences, if they exist, that distinguish the behaviour of CAH girls from that of their siblings. Cross-cultural study and a more sophisticated vocabulary for the description and classification of behaviour might help to avoid the barbarisms of ethnocentrism. Thus, it is at least theoretically possible that the description of data could be revised to minimize the biases of the investigators. We would then have a catalogue of human behaviour, dispositions, and behavioural differences that might or might not correspond to the socially salient distinctions of sex, race, and ethnicity. Perhaps we would also find physiological correlates for some of these differences. If this is indeed possible, then the masculine bias present in much behavioural description can be considered a function of inadequate analytic and descriptive tools and therefore incidental to the general project of developing a physiological account of behaviour and behavioural sex differences. Ironically, then, a feminist critique has the potential to improve and refine this area of inquiry.

Sexism does not seem intrinsic to the interpretation of data as evidence for physiological causal hypotheses. In our discussion of the distance between data and hypotheses and of the assumptions required to close that distance, we did, however, note that the assumption of cross-species uniformity and the adequacy of animal modelling is highly questionable in its application to behaviour. What explains its persistence, if not the role it plays in perpetuating the rule of difference, if *F* is present when *E* is absent, then *F* is unlikely to be a cause of *E*. Both presuppose the temporal priority of *F* to *E*. See John Stuart Mill, *A System of Logic*, 8th edn. (London: Longmans, Green & Co., 1949), 253–9.

18. Cf. Edward Wilson, *On Human Nature* (Cambridge, Mass.: Harvard University Press, 1978; New York: Bantam Press, 1979), 95.

19. We follow convention in distinguishing two forms of male bias. 'Androcentrism' applies to the perception of social life from a male point of view with a consequent failure accurately to perceive or describe the activity of women. 'Sexism' is reserved for statements, attitudes, and theories that presuppose, assert, or imply the inferiority of women, the legitimacy of their subordination, or the legitimacy of sex-based prescriptions of social roles and behaviours.

20. Barbara Fried, 'Boys Will Be Boys Will Be Boys', in Hubbard, Henifin, and Fried, (eds.) (n. 1 above), 37.

21. Donna Haraway, 'Animal Sociology and a Natural Economy of the Body Politic, pt. 2. The Past Is the Contested Zone: Human Nature and Theories of Production and Reproduction in Primate Behavior Studies', *Signs* 4/1 (Autumn 1978), 37–60, esp. 59.

22. Hubbard and Lowe (eds.) (n. 1 above), 27.

sexism? Historical and sociological analysis is required for a full answer to this question.

. . .

CONCLUSION

. . .

What constitutes an appropriate feminist response to masculine bias in science? Clearly this depends on the way bias is expressed in a given scientific context. Feminist anthropologists have developed alternative accounts of human evolution that replace androcentric with gynecentric assumptions while remaining within the methodological constraints of their disciplines. This strategy may not provide the final word in evolutionary theorizing, but it does reveal the epistemologically arbitrary nature of those androcentric assumptions and point the way to less restrictive understandings of human possibilities. As Donna Haraway has remarked regarding their work: 'The open future rests on a new past.'[21] Thus, one response is to adopt assumptions that are deliberately gynecentric or unbiased with respect to gender and see what happens.

In the case of the androcentric description of data, discerning masculine bias is only a first step. Questions remain regarding the phenomena shorn of tendentious description. Does androcentric language create or simply misdescribe its object? Some feminist critics have suggested that the entire category 'sex differences' is a fabrication supported by sexism and by analytic tendencies in science that emphasize distinctions over similarities.[22] More modestly, it can be argued that the concept 'tomboy' identifies but mystifies a slight difference in behaviour among young women. An alternative perspective might invent a name for young women who are not tomboys and seek the determinants of their peculiar behaviour. Scrutinizing the language used in the description of data can lead either to its disappearance as an object of inquiry or to the reformulation of the questions we ask about it.

When the issue concerns unreliable but not explicitly androcentric or sexist assumptions that are nevertheless suspected of being sexist in motivation, it is important not only to expose their unreliability but also to search for additional determinants. Such determinants may be embedded in the research programmes that grant these assumptions legitimacy, or they may be motivated by discriminatory intent other than sexism. Hereditarianism and various forms of biological determinism have been at the service of race and class supremacy as well as male domination. Because particular assumptions motivated by sexism are likely to be reinforced by additional types of bias in other contexts, they will not be dislodged by exposing their relation to sexism alone. Assumptions embedded in institutionalized research programmes offer a different challenge. Sometimes the critic will be able to show that their use in a given context is inappropriate. Other times she may have to be willing to take on the research project of an entire discipline.

As our methodological critique has shown, the variety of ways masculine bias expresses itself in science calls for—and permits—a variety of tactical responses. It is not necessary for us to turn our backs on science as a whole or to condemn it as an enterprise. In a number of ways, the logical structure of science itself provides opportunities for the expression of the creative and self-conscious sensibility that has characterized recent feminist attempts to transform the sciences.

Notes

1. See, e.g. Ruth Hubbard, Mary Henifin, and Barbara Fried (eds.), *Women Look at Biology Looking at Women* (Cambridge, Mass.: Schenkman Publishing Co., 1979); Ethel Tobach and Betty Rosoff (eds.), *Genes and Gender I* (New York: Gordian Press, 1977); Ruth Hubbard and Marian Lowe (eds.), *Genes and Gender II* (New York: Gordian Press, 1979); M. Kay Martin and Barbara Voorhies, *The Female of the Species* (New York: Columbia University Press, 1975); Evelyn Reed, *Sexism and Science* (New York: Pathfinder Press, 1978); and the special issue entitled 'Women, Science, and Society' of *Signs: Journal of Women in Culture and Society*, 4/1 (Autumn 1978).

2. Donna Haraway, 'In the Beginning Was the Word: The Genesis of Biological Theory', *Signs*, 6/3 (Spring 1981), 469–82, esp. 478.

3. Hubbard, Henifin, and Fried (eds.); Hubbard and Lowe (eds.). Our criticism of these authors' treatment of methodological issues is not meant to deny their contribution to the perception of sexually prejudicial aspects of science.

4. Hubbard and Lowe (eds.), 23–4, 144, 149; Ruth Hubbard, 'Have Only Men Evolved?' in Hubbard, Henifin, and Fried (eds.), 9.

5. Helen Longino, 'Evidence and Hypothesis', *Philosophy of Science* 46/1 (March 1979), 35–56.

6. The classic source for the man-the-hunter view is Sherwood Washburn and C. S. Lancaster, 'The Evolution of Hunting', in Richard Lee and Irven De Vore (eds.), *Man the Hunter* (Chicago: Aldine Publishing Co., 1968), 293–303. See also William Laughlin, 'Hunting: An Integrating Biobehavior System and Its Evolutionary Importance', in Lee and De Vore (eds.), 304–20. For woman the gatherer, see Nancy Tanner and Adrienne Zilhman, 'Women in Evolution, pt. 1. Innovation and Selection in Human Origins', *Signs* 1/3 (Spring 1976),

585–608; Adrienne Zihlman, 'Women in Evolution, pt. 2, Subsistence and Social Organization among Early Hominids', *Signs* 4/1 (Autumn 1978), 4–20; Nancy Tanner, *On Becoming Human* (Cambridge: Cambridge University Press, 1981); Adrienne Zihlman, 'Women as Shapers of the Human Adaptation', in Frances Dahlberg (ed.), *Woman the Gatherer* (New Haven, Conn.: Yale University Press, 1981), 75–120.

7. The analysis in the sections that follow is based on material derived from Gordon Bermant and Julian Davidson, *Biological Bases of Sexual Behavior* (New York: Harper & Row, 1974); Basil Eleftheriou and Richard Sprott (eds.), *Hormonal Correlates of Behavior* (New York: Plenum Press, 1975); Eleanor Maccoby and Carol Jacklin, *The Psychology of Sex Differences* (Stanford, Calif.: Stanford University Press, 1974); John Money and Anke Ehrhardt, *Man and Woman, Boy and Girl* (Baltimore: Johns Hopkins University Press, 1972); Kenneth Moyer (ed.), *The Physiology of Aggression* (New York: Raven Press, 1976); Susan Baker, 'Biological Influences on Sex and Gender', *Signs* 6/1 (Autumn 1980), 80–96; and the review articles on the biological bases of sex differences in *Science*, 211 (20 Mar. 1981), 1265–1324.

8. Robert Goy and Bruce McEwen, *Sexual Differentiation of the Brain* (Cambridge, Mass.: MIT Press, 1980), 79.

9. Robert Goy and David Goldfoot, 'Neuroendocrinology: Animal Models and Problems of Human Sexuality', in Eli Rubinstein and Richard Green (eds.), *New Directions in Sex Research* (New York: Plenum Press, 1975), 83–98. For criticism of such views, see Julian Davidson, 'Biological Models of Sex: Their Scope and Limitations', in Herant Katchadourian (ed.), *Human Sexuality* (Berkeley and Los Angeles: University of California Press, 1979), 134–49.

10. Stephen Goldberg, *The Inevitability of Patriarchy* (New York: William Morrow & Co., 1973).

11. Maccoby and Jacklin, 243–7.

12. Ibid. 263–5, 274, 368–71.

13. Anke Ehrhardt and Heino Meyer-Bahlburg, 'Effects of Prenatal Sex Hormones on Gender-related Behavior', *Science*, 211 (20 Mar. 1981), 1312–18.

14. For this representation of Ehrhardt's views we rely primarily on Ehrhardt and Meyer-Bahlburg. Ehrhardt's earlier publications on this subject have been critically discussed by Elizabeth Adkins, 'Genes, Hormones, Sex and Gender', in *Sociobiology: Beyond Nature/Nurture?* ed. George Barlow and James Silverberg (Washington, DC: American Association for the Advancement of Science, 1980), 385–415.

15. Margaret Mead, *Sex and Temperament in Three Primitive Societies* (New York: William Morrow & Co., 1935), is still an excellent source for this point of view. See also Beatrice Whiting and Carolyn Pope Edwards, 'A Cross-cultural Analysis of Sex Differences in the Behavior of Children Aged Three to Eleven', *Journal of Social Psychology*, 91 (1973), 171–88.

16. Ehrhardt (see Ehrhardt and Meyer-Bahlburg) notes that the only influence acknowledged by the parents was encouragement of 'feminine' rather than 'masculine' behaviour. Self-reporting is not, however, the most reliable source of information in a sensitive area like child-rearing. In addition, effective parental influence is rarely overt or conscious.

17. The rule of agreement states that if *F* is present whenever *E* is present and absent whenever *E* is absent, then *F* is likely to be a cause of *E*. According to

6 Pre-Theoretical Assumptions in Evolutionary Explanations of Female Sexuality

Elisabeth A. Lloyd

. . .

In the second wave of the feminist movement in the United States, debates about women's identity have explicitly included sexuality; much feminist argument in the late 1960s and early 1970s involved an attempt to separate out an autonomous female sexuality from women's reproductive functions.

It is especially relevant, then, to examine biological arguments, particularly evolutionary arguments, to see what they say about *whether* and *how* women's sexuality is related to reproduction. We shall find that many evolutionary arguments seem to support the direct linking of female sexuality and reproduction. Yet I will argue that this support is not well-grounded. In fact, I think evolutionary explanations of female sexuality exemplify how social beliefs and social agendas can influence very *basic* biological explanations of fundamental physiological processes. In this paper, I shall spend some time spelling out a few examples in which assumptions about the close link between reproduction and sexuality yield misleading results, then I shall conclude with a discussion of the consequences of this case-study for issues in the philosophy of science.

The fundamental problem is that it is simply *assumed* that every aspect of female sexuality should be explained in terms of reproductive functions. But there is quite a bit of biological evidence that this is an empirically incorrect assumption to make. This raises the question of why autonomous female sexuality, distinct from reproductive functions, got left out of these explanations. I shall ultimately conclude that social context is playing a large and unacknowledged role in the practice of this science.

Perhaps the notion of the potential independence of female sexuality and reproduction may be unclear: I suggest thinking in terms of two distinct models, one in which all basic aspects of sexuality are

Reprinted with permission from *Philosophical Studies*, 69 (1993), 139–53.

explained in terms of reproduction, and the other in which sexuality is seen as an autonomous set of functions and activities, which are only *partially* explained in terms of reproductive functions. The difference may seem minor, but the two models have significantly disparate consequences when used in scientific explanation.

Let us begin with a classic and widespread model representing the hormonal determination of sexual behaviour. In this model, female animals are only willing to have sexual intercourse when they are fertile—their sexual interest and activity are completely hormonally controlled. Typical and familiar examples of this type of set-up include rats, dogs, and cats. When these animals are in oestrus, they are willing and eager to mate, otherwise not. Technically, oestrus is defined hormonally—that is, oestrus is a particular phase of the menstrual cycle, in which the animal is fertile, and certain hormone measures are very high. This model embodies a *very tight* link between sexuality and reproduction: female sexuality functions completely in the service of reproduction.

Some interesting problems arise, however, in the application of this hormonally deterministic picture to human and non-human primate behaviour.

First of all, although oestrus is biologically defined as a hormonal state, it is very common for oestrus to be defined *operationally* as the period in which 'the female is willing to participate in sex'. In one species, the bonobos, this behavioural definition led to the comic conclusion that this species is in oestrus 57–86 per cent of the time.[1] Notice that identifying oestrus in this manner amounts to an *enforcement* of the belief that sexual behaviour is tightly linked to reproduction. It becomes impossible even to *ask* whether these primates have an active sexual interest outside of their peak hormonal periods.

It turns out that when independent studies are made, sexual activity is not confined to the fertile phase for a number of non-human primates, including rhesus monkeys, several species of baboons, and common chimpanzees.[2] Social factors such as partner preferences can be as influential as hormonal factors in regulating sexual behaviour in several of these species.[3]

Female homosexual activity provides a good test for the assumed dependence of female sexuality on hormonal status. In addition, homosexual behaviours are clearly independent of reproduction *per se*, and might be interpreted as an indicator of an autonomous female sexuality. It turns out that female homosexual activities, which are widely observed in non-human primates, seem to be *independent* of the hormonal status of the participants. This independence has led some researchers to ignore such behaviours, or to declare that they are not, in fact, sexual. For example, pygmy chimpanzee females are commonly observed engaging in 'genito-genital rubbing' (called 'GG rubbing') in which two females hold each other and 'swing their hips laterally while keeping the front tips of vulvae, where the clitorises protrude, in touch with each other'.[4] Kano argues that this behaviour is not sexual, because non-human primates can only be 'sexual' during oestrus; the fact that pygmy chimps engage in GG rubbing outside of oestrus, claims Kano, itself 'suggests that this behaviour does not occur exclusively in a sexual context, but has some other social significance'.[5] Generally, some caution about the interpretation of apparently sexual behaviours is appropriate; the misunderstanding of many dominance behaviours as sexual ones plagued primatology in its first decades. At stake in this case, however, is the very *possibility* of hormonally independent female sexuality. The issue was resolved in 1984, when Mori, using a detailed study of statistical relations among behaviours, concluded that GG rubbing was, in fact, sexual behaviour, since the same cluster of behaviour surrounded both mating and GG rubbing.[6]

A more blatant example of researcher-bias typing reproduction and female sexuality tightly together appears in an experiment being done on female orgasm in stumptail macaques. The original studies on female macaque orgasm, completed in the 1970s, documented female orgasm in the context of female homosexual mounting—that is, one female mounts another female, and stimulates herself to orgasm.[7] One very interesting result of these studies was the finding that the mounting, orgasmic female was *never* in oestrus when these orgasms occurred. This is a provocative result for several reasons. First, according to the hormonal determinism model, female macaques are not supposed to be interested in any sexual activity outside of oestrus; Second, these same female macaques *never* evidenced any sign of orgasm when they were participating in heterosexual coitus. A later study of the same species documented the same basic patterns, with the exception that four out of ten females in the group seemed, occasionally, to have orgasm during heterosexual coitus.[8]

I was surprised, therefore, when I spoke with a researcher who was working on the evolution of female orgasm in stumptail macaques.[9] He described his experimental set-up to me with some enthusiasm: the females are radio-wired to record orgasmic muscle contractions and increased heart rate, etc. This sounds like the ideal experiment, because it can record the sex lives of the females mechanically, without needing a human observer. In fact, the project had been funded by the NIH, and had presumably gone through the outside referee and panel reviews necessary for funding. But then the researcher described to me the clever way he had set up his equipment to record the female orgasms—he wired up the heart rate of the *male* macaques as the signal to start recording the *female* orgasms. When I pointed out that the vast majority of female stumptail orgasms occurred during sex among the females alone, he replied that yes, he knew that, but he was only interested in the *important* orgasms.

Obviously, this is a very unfortunate case. But it is not an isolated incident. Observations, measurements, interpretations, and experimental design are all affected by the background assumptions of the scientists. There is a pervasive and undefended assumption that female sexuality in non-human primates is tightly linked to reproduction. I would like now to explore briefly the situation regarding human beings.

HUMAN CASES

In most of the literature on the evolution of human sexuality, much attention is paid to the distinct attributes of human beings. The continual sexual 'receptivity' of the human female is contrasted with the (supposed) strict hormonal restrictions on sexual activity in non-human animals. Human beings are supposed to be uniquely adapted to be sexually free from hormonal dictates, the possessors of a separate and self-constructed sexuality. When it comes to evolutionary explanations of women's sexuality, though, the tight connection between reproduction and sexuality remains firmly in place. . . .

Many researchers, in evolutionary biology, behaviour, and physiology, have *deduced* that it must be the case in human females that peak sexual interest and desire occur at the same time as peak fertility. This conclusion is a simple extension of the hormonal determinism model from mice and dogs. While this may have the ring of a reasonable assumption, it is not supported by the clinical literature. Kinsey, for example, found that 59 per cent of his female sample experienced patterns of fluctuation in their sexual desire during their cycle—but only 11 per cent experience a peak of sexual desire in mid-cycle, when they are most likely to be fertile.[10] More recently, Singer and Singer, in a survey of studies, found that only 6–8 per cent of women experience an increase in sexual desire around the time of ovulation. Most studies found peaks of sexual desire right before and after menstruation, when the woman is almost invariably infertile.[11]

. . .

Unfortunately, a number of researchers working in the area of the evolution of sexuality have not taken this on board, and continue to assert that peak sexual desire *must* be around the time of ovulation—otherwise it would not make any sense.

This 'making sense' is precisely what I'm interested in. According to these researchers, female sexuality doesn't *make sense* unless it is in the service of reproduction. There is no scientific defence offered for this assumption. A similar assumption is also present in the evolutionary explanations offered for female orgasm.

I have examined thirteen stories for the evolution of human female orgasm, and all except one of these stories assume that orgasm is an evolutionary adaptation. That is, they assume that orgasm conferred a *direct selective* advantage on its possessors, and that is how it came to be prevalent among women. The most common general formula for explaining the evolution of human female orgasm is through the pair bond. Here, the pair bond means more-or-less monogamous heterosexual coupling, and it is argued that such coupling increases the potential reproductive success of both parties through mutual co-operation and assistance with rearing offspring. The idea is that the male and the female in the pair bond provide mutual support to one another, and assist each other in rearing offspring, and that offspring raised under these conditions will tend themselves to have higher survival and reproductive success than those raised under other circumstances.

Hence, pair bonding is seen as an adaptation in the evolutionary sense—it exists *because* it confers better chances of surviving and reproducing to those who display the trait. Under the assumption that pair bonds are adaptive, frequent intercourse is also seen as adaptive, since it helps 'cement the pair bond'. And this is where orgasm comes in. Orgasm evolved, according to these pair bond theorists, because it gave the female a reward and motivation to engage in frequent intercourse, which is itself adaptive, because it helps cement the pair bond. A number of different theorists have developed permutations of this basic story, but it remains the most widely accepted evolutionary story for female orgasm.[12]

Now, there is a glaring problem with this story—it assumes that intercourse is reliably connected to orgasm in females. All of the available clinical studies on women's sexual response indicate that this is a problematic assumption. Somewhere between 20–35 per cent of women always or almost always experience orgasm with unassisted intercourse.[13] I should add that this figure is supported by what cross-cultural information exists.[14] This figure is very low, and it is especially striking given that somewhere around 90 per cent of women do experience orgasm. Furthermore, about 30 per of women *never* have orgasm with intercourse—this figure is taken from a population of women who do have regular intercourse, and of whom almost all are orgasmic.[15] What this means is that *not* to have orgasm from intercourse is the experience of the majority of women the majority of the time. Not to put too fine a point on it, if orgasm is an adaptation which is a reward for engaging in frequent intercourse, it does not seem to work very well.

Obviously, this observation does not rule out the possibility that there is some selective advantage to female orgasm, but the salient point is that *none of these pair bond theorists even address this problem*, which I call the orgasm–intercourse discrepancy. Rather they simply assume that when intercourse occurs, so does orgasm.[16]

In general, the association of intercourse with orgasm is relatively unproblematic among males. Hence, what is being assumed here is that female sexual response is like male sexual response to the same situation. There is little or no awareness, among the pair bond theorists, of the orgasm–intercourse discrepancy, in spite of the fact that they cite or refer to the very studies which document this fact, including Kinsey's 1953 report on women's sexual response.

There is one obvious and understandable reason for this slip. They are, after all, trying to explain orgasm through evolutionary theory, which involves showing that the trait gave a reproductive advantage to its owner. It's easy to see how the equation of reproduction through intercourse and orgasm went by unnoticed. Nevertheless, this case does illustrate the main thesis, that female sexuality is unquestioningly equated with reproduction, and with the sort of sex that leads to reproduction.

There is another intriguing line of argument for the adaptive value of female orgasm, which was first published by Desmond Morris in 1967, though Shirley Strum tells me that Sherwood Washburn was teaching this in his classes at Berkeley earlier. Morris claimed that orgasm had a special function related to bipedalism (that is, walking on our hind legs), because it would increase chances of fertilization. . . .

There is . . . a great advantage in any reaction that tends to keep the female horizontal when the male ejaculates and stops copulation. The violent response of female orgasm, leaving the female sexually satiated and exhausted, has precisely this effect.[17]

Morris' view is in turn based on his understanding of physiological response—he says earlier '. . . after both partners have experienced orgasm [in intercourse] there normally follows a considerable period of exhaustion, relaxation, rest and frequently sleep'.[18] Similarly, he claims, 'once the climax has been reached, all the [physiological] changes noted are rapidly reversed and the resting, post-sexual individual quickly returns to the normal quiescent physiological state.'[19]

Now let us refer to the clinical sex literature, which is cited by Morris and by others. According to this literature, the tendencies to states of sleepiness and exhaustion following orgasm, are, in fact, true for men but not for women. Regarding Morris's claim that the physiological changes are 'rapidly reversed', this is also true for men but not for women—women return to the plateau phase of sexual excitement, and not to the original unexcited phase, as men do. This was one of the most noted conclusions of Masters and Johnson, whose picture of sexual response was enthusiastically adopted by Morris—but, it seems, only in part.[20]
. . .

Actually, another serious problem with this story was recently pointed out by Shirley Strum, an expert on baboon behaviour.[21] Supposedly, the selection pressure shaping female sexual response here is the potential loss of sperm that is threatened because human beings walk on two legs, and because the vaginal position is thus changed from horizontal to almost vertical. One would think, then, that our relatives walking on four legs would be protected against this occurrence, for anatomical reasons. But Strum says that immediately following intercourse, female baboons like to go off and *sit down* for ten or fifteen minutes. When they get up, she says, they inevitably leave a visible puddle of semen on the ground. Perhaps, then, the loss of semen is not the serious evolutionary challenge that Desmond Morris and others take it to be.

SUMMARY

I claim that social agendas appear in these stories through the obliteration of any female sexual response that is independent from her function as a reproducer. Autonomous, distinct female sexual response just disappears.

In these explanations women are presumed to have orgasms nearly always with intercourse, as men do. Women are presumed to return to the resting state following orgasm, as men do. One could object that Morris is a relatively easy target, so I will offer the following titbit in defence of my analysis. Gordon Gallup and Susan Suarez published, in 1983, a technical discussion on optimal reproductive strategies for bipedalism, and took up Morris' anti-gravity line of argument. They argue that orgasm would be adaptive because it would keep the woman lying down, and hence keep the semen from escaping. In the context of these paragraphs on female orgasm, they state, 'it is widely acknowledged that intercourse frequently acts as a mild sedative. The average individual requires about five minutes of repose before returning to a normal state after orgasm.'[22] The scientific reference they offer for this particular generalization is Kinsey 1948, which is, in fact, exclusively on *male* sexual response. In other words, this 'average individual' which figures in their story about female orgasm, is, in fact, explicitly male.

AN ALTERNATIVE EXPLANATION

Donald Symons, in his book *The Evolution of Human Sexuality* (1979), argues that female orgasm is not an adaptation. He develops a story parallel to the one about male nipples—female orgasm exists because orgasm is strongly selected in males, and because of their common embryological form, women are born with the potential for having orgasms, too.[23] Part of the story, then, is that orgasm is strongly selected in males; this is fairly plausible, since it is difficult for male mammals to reproduce without ejaculation, which requires a reflex response in certain muscles. These muscles are, in fact, the same (homologous) muscles that are involved in female orgasm. It is also significant that the intervals between contractions in orgasm is four-fifths of a second in both men and women. This is considered evidence that orgasm is a reflex with the same developmental origin in both sexes.

One of the consequences of Symons' theory is that it would be expected that similar stimulation of the clitoris and penis would be required to achieve the same reaction or reflex response. This similarity shows especially in the figures on masturbation. Only 1.5 per cent of women masturbate by vaginal entry, which provides stimulation similar to the act of intercourse; the rest do so by direct or indirect stimulation of the clitoris itself.[24] Also, on the developmental theory, one would *not* expect similar reactions to intercourse, given the differences in stimulation of the homologous organs.

Finally, this theory is also supported by the evidence of orgasm in non-human primates. The observed orgasms occur almost exclusively when the female monkeys are themselves mounting other monkeys, and not during copulation. On the non-adaptive view of orgasm, this is almost to be expected. There, female orgasm is defined as a potential, which, if the female gets the right sort and amount of stimulation, is activated. Hence, it is not at all surprising that this does not occur often during copulation, which in these monkeys includes very little, if any, stimulation of the clitoris, but occurs rather with analogous stimulation of the homologous organs that they get in mounting.

. . .

The conclusion that orgasm is not an adaptation *could* be interpreted as emancipatory. After all, the message here is that orgasm is a freebie. It can be used in any way that people want; there is no 'natural' restriction on female sexual activities, nor is there any scientific ground for such a notion. Under the developmental view, the constraints are loosened on possible explanations about women's sexuality that are consistent with accepted clinical conclusions and with evolutionary theory. Hence, the realm formerly belonging exclusively to reproductive drive would now be open to much, much more.

DISCUSSION

I would like to draw two conclusions.

First, I believe that prior assumptions have more influence in these areas of science than is commonly acknowledged in the usual philosophical and scientific pictures of scientific theorizing and testing. In the cases examined here, science is not very separate from the social and cultural context. Rather, social assumptions and prior commitments of the scientists play a major role in the practice of science itself, at many levels—experimental design, data collection, predictions, hypothesis formulation, and the evaluation of explanations.

. . .

My suggestion does *not* involve commitment to a relativist position. In a complete analysis of evolutionary explanations of human sexuality, I would adopt Helen Longino's general approach, in which she characterizes objectivity in science as resulting from the critical interaction of different groups and individuals with different social and cultural assumptions and different stakes. Under this view, the irreducibility of the social components of the scientific situation is accounted for—these social assumptions are, in fact, an essential part of the picture of scientific practice.

At any rate, I take it that the cases I have described above violate our common philosophical understandings of how we arrive at scientific beliefs, how knowledge is created, and how science works. If philosophers go the route of labelling as 'science' *only* that which obeys the demands of current philosophy, we will end up discussing only some parts of physics and maybe some maths. Meanwhile, what about the rest of science—biology, social sciences, anthropology, psychology, biochemistry? I suggest adopting and developing recent contextualist and feminist views of science, which take explicit account of pre-theoretical assumptions and preconceptions, and their social origins.

This case involving female sexuality is very interesting because there are *two* very strong forces working to put sex and reproduction together. Adaptationism, within biology, promotes the easy linking of all sexual activity with reproduction success, the measure of relative fitness. Secondly, the long social tradition of *defining* women in terms of their sexual and reproductive functions alone also tends to link sexuality and reproduction more tightly than the evidence indicates.

The long struggle by various women's movements to separate sex and reproduction seems to have had very little effect on the practice of the science we have examined in this paper. This is especially ironic, because politically, ever since the late nineteenth century, scientific views about gender differences and the biology of women have been the single most powerful political tool against the women's movements. My second and more controversial conclusion is that current 'purist' philosophy of science actually *contributes to* that political power by reinforcing myths of the insulation of scientific endeavours from social influences. A more sophisticated understanding of the production and evaluation of scientific knowledge would mean seeing science as (partly) a continuation of politics. Science would then lose at least *some* independent authority in the political arena. Judging by the scientific work that I have discussed in this paper, I think that would be a good thing.

Notes

1. Kano, T. (1982), 'The Social Group of Pygmy Chimpanzees of Wamba', *Primates*, 23/2: 171–88.
2. Hafez, E. S. E. (1971), 'Reproductive Cycles', in *Comparative Reproduction of Non-Human Primates*, ed. E. S. E. Hafez (Springfield, Ill.: Charles C. Thomas).
3. Wolfe, L. (1979), 'Behavioral Patterns of Estrous Females of the Arachiyama West Troop of Japanese Macaques (*Macaca fuscata*)', *Primates*, 20/4: 525–34.
4. Koruda, S. (1980), 'Social Behavior of the Pygmy Chimpanzees', *Primates*, 21/2: 181–97. Quote from p. 189.
5. Kano, T. (1980), 'Social Behavior of Wild Pygmy Chimpanzees (*Pan paniscus*) of Wambe: A Preliminary Report', *Journal of Human Evolution*, 9: 243–60. Quote from p. 243.
6. Mori, A. (1984), 'An Ethological Study of Pygmy Chimpanzees in Wambe, Zaire: A Comparison with Chimpanzees', *Primates*, 25/3: 255–78.
7. Chevalier-Skolnikoff, S. (1974), 'Male–Female, Female–Female, and Male–Male Sexual Behavior in the Stumptail Monkey, with Special Attention to the Female Orgasm', *Archives of Sexual Behavior*, 3/2: 95–116; (1976) 'Homosexual Behavior in a Laboratory Group of Stumptail Monkeys (*Macaca arctoides*): Forms, Contexts, and Possible Social Functions', *Archives of Sexual Behavior*, 5/6: 511–27.
8. Goldfoot, D., J. Westerborg-van Loon, W. Groeneveld, and A. Koos Slob (1980), 'Behavioral and Physiological Evidence of Sexual Climax in the Female Stump-Tailed Macaque (*Macaca arctoides*)', *Science*, 208: 1477–9.
9. The identity of this researcher is not included for publication. The information stated here was obtained through personal communication.
10. Kinsey, A. C. *et al.* (1953), *Sexual Behavior in the Human Female* (Philadelphia: W. B. Saunders).
11. Singer, I. and J. Singer (1972), 'Periodicity of Sexual Desire in Relation to Time of Ovulation in Women', *Journal of Biosocial Science*, 4: 471–81.
12. Morris' work has been criticized by later researchers as being methodologically flawed (e.g. E. O. Wilson (1975), *Sociobiology* (Harvard University Press); J. H. Crook (1972), 'Sexual Selection, Dimorphism, and Social Organization in the Primates', in B. Campbell (ed.), *Sexual Selection and the Descent of Man* (Chicago: Aldine) but it is still widely cited, and its basic premises are accepted or slightly modified in other respected accounts such as: F. Beach (1973), 'Human Sexuality and Evolution', in W. Montanga and W. Sadler (eds.),

Advances in Behavioral Biology (New York: Plenum Press), 333–65; G. Pugh (1977), *Biological Origins of Human Values* (New York: Basic Books); Crook (1972); and B. Campbell (1967), *Human Evolution: An Introduction to Man's Adaptations* (Chicago: Aldine).

13. Hite, S. (1976), *The Hite Report* (New York: Macmillan); Kinsey, A. *et al.* op. cit.

14. See, e.g. Davenport, W. (1977), 'Sex in Cross-Cultural Perspective', in F. Beach (ed.), *Human Sexuality in Four Perspectives* (Johns Hopkins University Press), 115–63.

15. Hite, S., op. cit.

16. Typically, in evolutionary explanations, if a trait is taken to have evolved as an adaptation, yet is rarely used in the adaptive context, some explanation of the details of the selection pressure or the extreme adaptive value of the trait is offered.

17. Morris, D. (1967), *The Naked Ape* (London: Jonathan Cape), 79.

18. Ibid. 55.

19. Ibid. 59.

20. Masters, W. H. and V. Johnson (1966), *Human Sexual Response* (Boston: Little, Brown).

21. Personal communication.

22. Gallup, G. and S. Suarez (1983), 'Optimal Reproductive Strategies for Bipedalism', *Journal of Human Evolution*, 12: 195.

23. This argument is spelled out in more detail by Stephen Jay Gould, in an essay that was based on my research and arguments ('Freudian Slip', *Natural History* (Feb. 1987), 14-21).

24. Kinsey, 1953; Hite (1976), 410–11.

7 The Egg and the Sperm: How Science has Constructed a Romance Based on Stereotypical Male–Female Roles

Emily Martin

The theory of the human body is always a part of a world-picture. . . . The theory of the human body is always a part of a *fantasy*.[1]

As an anthropologist, I am intrigued by the possibility that culture shapes how biological scientists describe what they discover about the natural world. If this were so, we would be learning about more than the natural world in high school biology class; we would be learning about cultural beliefs and practices as if they were part of nature. In the course of my research I realized that the picture of egg and sperm drawn in popular as well as scientific accounts of reproductive biology relies on stereotypes central to our cultural definitions of male and female. The stereotypes imply not only that female biological processes are less worthy than their male counterparts but also that women are less worthy than men. Part of my goal in writing this article is to shine a bright light on the gender stereotypes hidden within the scientific language of biology. Exposed in such a light, I hope they will lose much of their power to harm us.

EGG AND SPERM: A SCIENTIFIC FAIRY TALE

At a fundamental level, all major scientific textbooks depict male and female reproductive organs as systems for the production of valuable substances, such as eggs and sperm.[2] In the case of women, the monthly cycle is described as being designed to produce eggs and prepare a suitable place for them to be fertilized and grown—

Reprinted from *Signs: Journal of Women in Culture and Society*, 16/3 (1991). By permission of the University of Chicago Press.

all to the end of making babies. But the enthusiasm ends there. By extolling the female cycle as a productive enterprise, menstruation must necessarily be viewed as a failure. Medical texts describe menstruation as the 'debris' of the uterine lining, the result of necrosis, or death of tissue. The descriptions imply that a system has gone awry, making products of no use, not to specification, unsaleable, wasted, scrap. An illustration in a widely used medical text shows menstruation as a chaotic disintegration of form, complementing the many texts that describe it as 'ceasing', 'dying', 'losing', 'denuding', 'expelling'.[3]

Male reproductive physiology is evaluated quite differently. One of the texts that sees menstruation as failed production employs a sort of breathless prose when it describes the maturation of sperm: 'The mechanisms which guide the remarkable cellular transformation from spermatid to mature sperm remain uncertain. . . . Perhaps the most amazing characteristic of spermatogenesis is its sheer magnitude: the normal human male may manufacture several hundred million sperm per day.'[4] In the classic text *Medical Physiology*, edited by Vernon Mountcastle, the male–female, productive–destructive comparison is more explicit: 'Whereas the female *sheds* only a single gamete each month, the seminiferous tubules *produce* hundreds of millions of sperm each day' (emphasis mine).[5] The female author of another text marvels at the length of the microscopic seminiferous tubules, which, if uncoiled and placed end to end, 'would span almost one-third of a mile!' She writes, 'In an adult male these structures produce millions of sperm cells each day.' Later she asks, 'How is this feat accomplished?'[6] None of these texts expresses such intense enthusiasm for any female processes. It is surely no accident that the 'remarkable' process of making sperm involves precisely what, in the medical view, menstruation does not: production of something deemed valuable.[7]

One could argue that menstruation and spermatogenesis are not analogous processes and, therefore, should not be expected to elicit the same kind of response. The proper female analogy to spermatogenesis, biologically, is ovulation. Yet ovulation does not merit enthusiasm in these texts either. Textbook descriptions stress that all of the ovarian follicles containing ova are already present at birth. Far from being *produced*, as sperm are, they merely sit on the shelf, slowly degenerating and aging like overstocked inventory: 'At birth, normal human ovaries contain an estimated one million follicles [each], and no new ones appear after birth. Thus, in marked contrast to the male, the newborn female already has all the germ cells she will ever have. Only a few, perhaps 400, are destined to reach full maturity during her active productive life. All the others degenerate at some point in their development so that few, if any, remain by the time she reaches menopause at approximately 50 years of age.'[8] Note the 'marked contrast' that this description sets up between male and female: the male, who continuously produces fresh germ cells, and the female, who has stockpiled germ cells by birth and is faced with their degeneration.

Nor are the female organs spared such vivid descriptions. One scientist writes in a newspaper article that a woman's ovaries become old and worn out from ripening eggs every month, even though the woman herself is still relatively young: 'When you look through a laparoscope . . . at an ovary that has been through hundreds of cycles, even in a superbly healthy American female, you see a scarred, battered organ.'[9]

To avoid the negative connotations that some people associate with the female reproductive system, scientists could begin to describe male and female processes as homologous. They might credit females with 'producing' mature ova one at a time, as they're needed each month, and describe males as having to face problems of degenerating germ cells. This degeneration would occur throughout life among spermatogonia, the undifferentiated germ cells in the testes that are the long-lived, dormant precursors of sperm.

. . .

The real mystery is why the male's vast production of sperm is not seen as wasteful.[10] Assuming that a man 'produces' 100 million (10^8) sperm per day (a conservative estimate) during an average reproductive life of sixty years, he would produce well over two trillion sperm in his lifetime. Assuming that a woman 'ripens' one egg per lunar month, or thirteen per year, over the course of her forty-year reproductive life, she would total five hundred eggs in her lifetime. But the word 'waste' implies an excess, too much produced. Assuming two or three offspring, for every baby a woman produces, she wastes only around two hundred eggs. For every baby a man produces, he wastes more than one trillion (10^{12}) sperm.

How is it that positive images are denied to the bodies of women? A look at language—in this case, scientific language—provides the first clue. Take the egg and the sperm. It is remarkable how 'femininely' the egg behaves and how 'masculinely' the sperm.[11] The egg is seen as large and passive.[12] It does not *move* or *journey*, but passively 'is transported', 'is swept',[13] or even 'drifts'[14] along the fallopian tube. In utter contrast, sperm are small, 'streamlined',[15] and invariably active. They 'deliver' their genes to the egg, 'activate the developmental program of the egg',[16] and have a 'velocity' that is often remarked upon.[17] Their tails are 'strong' and efficiently powered.[18] Together with the forces of ejaculation, they can 'propel the semen into the deepest recesses of the vagina'.[19] For this they need 'energy',[20] 'fuel',[20] so that with a 'whiplashlike motion and strong lurches'[21] they can 'burrow through the egg coat'[22] and 'penetrate' it.[23]

At its extreme, the age-old relationship of the egg and the sperm takes on a royal or religious patina. The egg coat, its protective barrier, is sometimes called its 'vestments', a term usually reserved for sacred, religious dress. The egg is said to have a 'corona',[24] a crown, and to be accompanied by 'attendant cells'.[25] It is holy, set apart and above, the queen to the sperm's king. The egg is also passive, which means it must depend on sperm for rescue. Gerald Schatten and Helen Schatten liken the egg's role to that of Sleeping Beauty: 'a dormant bride awaiting her mate's magic kiss, which instills the spirit that brings her to life'.[26] Sperm, by contrast, have a 'mission',[27] which is to 'move through the female genital tract in quest of the ovum'.[28] One popular account has it that the sperm carry out a 'perilous journey' into the 'warm darkness', where some fall away 'exhausted'. 'Survivors' 'assault' the egg, the successful candidates 'surrounding the prize'.[29] Part of the urgency of this journey, in more scientific terms, is that 'once released from the supportive environment of the ovary, an egg will die within hours unless rescued by a sperm'.[30] The wording stresses the fragility and dependency of the egg, even though the same text acknowledges elsewhere that sperm also live for only a few hours.[31]

In 1948, in a book remarkable for its early insights into these matters, Ruth Herschberger argued that female reproductive organs are seen as biologically interdependent, while male organs are viewed as autonomous, operating independently and in isolation:

. . .

The sperm is no more independent of its milieu than the egg, and yet from a wish that it were, biologists have lent their support to the notion that the human female, beginning with the egg, is congenitally more dependent than the male.[32]

An article in the journal *Cell* has the sperm making an 'existential decision' to penetrate the egg: 'Sperm are cells with a limited behavioral repertoire, one that is directed toward fertilizing eggs. To execute the decision to abandon the haploid state, sperm swim to an egg and there acquire the ability to effect membrane fusion.[33] Is this a corporate manager's version of the sperm's activities—'executing decisions' while fraught with dismay over difficult options that bring with them very high risk.

. . .

One depiction of sperm as weak and timid, instead of strong and powerful—the only such representation in Western civilization, so far as I know—occurs in Woody Allen's movie *Everything You Always Wanted To Know About Sex* *But Were Afraid to Ask*. Allen, playing the part of an apprehensive sperm inside a man's testicles, is scared of the man's approaching orgasm. He is reluctant to launch himself into the darkness, afraid of contraceptive devices, afraid of winding up on the ceiling if the man masturbates.

The more common picture—egg as damsel in distress, shielded only by her sacred garments; sperm as heroic warrior to the rescue—cannot be proved to be dictated by the biology of these events. While the 'facts' of biology may not *always* be constructed in cultural terms, I would argue that in this case they are. The degree of metaphorical content in these descriptions, the extent to which differences between egg and sperm are emphasized, and the parallels between cultural stereotypes of male and female behaviour and the character of egg and sperm all point to this conclusion.

NEW RESEARCH, OLD IMAGERY

As new understandings of egg and sperm emerge, textbook gender imagery is being revised. But the new research, far from escaping the stereotypical representations of egg and sperm, simply replicates elements of textbook gender imagery in a different form. The persistence of this imagery calls to mind what Ludwik Fleck termed 'the self-contained' nature of scientific thought. As he described it, 'the interaction between what is already known, what remains to be learned, and those who are to apprehend it, go to ensure harmony within the system. But at the same time they also preserve the harmony of illusions, which is quite secure within the confines of a given thought style.[34] We need to understand the way in which the cultural content in scientific descriptions changes as biological discoveries unfold, and whether that cultural content is solidly entrenched or easily changed.

In all of the texts quoted above, sperm are described as penetrating the egg, and specific substances on a sperm's head are described as binding to the egg. Recently, this description of events was rewritten in a biophysics lab at Johns Hopkins University—transforming the egg from the passive to the active party.[35]

Prior to this research, it was thought that the zona, the inner vestments of the egg, formed an impenetrable barrier. Sperm overcame the barrier by mechanically burrowing through, thrashing their tails and slowly working their way along. Later research showed that the sperm released digestive enzymes that chemically broke down the zona; thus, scientists presumed that the sperm used mechanical *and* chemical means to get through to the egg.

In this recent investigation, the researchers began to ask questions about the mechanical force of the sperm's tail. (The lab's goal was to develop a contraceptive that worked topically on sperm.) They discovered, to their great surprise, that the forward thrust of sperm is extremely weak, which contradicts the assumption that sperm are forceful penetrators.[36] Rather than thrusting forward, the sperm's head was now seen to move mostly back and forth. The sideways motion of the sperm's tail makes the head move sideways with a force that is ten times stronger than its forward movement. So even if the overall force of the sperm were strong enough to mechanically break the zona, most of its force would be directed sideways rather than forward. In fact, its strongest tendency, by tenfold, is to escape by attempting to pry itself off the egg. Sperm, then, must be exceptionally efficient at *escaping* from any cell surface they contact. And the surface of the egg must be designed to trap the sperm and prevent their escape. Otherwise, few if any sperm would reach the egg.

The researchers at Johns Hopkins concluded that the sperm and egg stick together because of adhesive molecules on the surfaces of each. The egg traps the sperm and adheres to it so tightly that the sperm's head is forced to lie flat against the surface of the zona, a little bit, they told me, 'like Br'er Rabbit getting more and more stuck to tar baby the more he wriggles'. The trapped sperm continues to wiggle ineffectually side to side. The mechanical force of its tail is so weak that a sperm cannot break even one chemical bond. This is where the digestive enzymes released by the sperm come in. If they start to soften the zona just at the tip of the sperm and the sides remain stuck, then the weak, flailing sperm can get oriented in the right direction and make it through the zona—provided that its bonds to the zona dissolve as it moves in.

Although this new version of the saga of the egg and the sperm broke through cultural expectations, the researchers who made the discovery continued to write papers and abstracts as if the sperm were the active party who attacks, binds, penetrates, and enters the egg. The only difference was that sperm were now seen as performing these actions weakly.[37] Not until August 1987, more than three years after the findings described above, did these researchers reconceptualize the process to give the egg a more active role. They began to describe the zona as an aggressive sperm catcher, covered with adhesive molecules that can capture a sperm with a single bond and clasp it to the zona's surface.[38] In the words of their published account: 'The innermost vestment, the *zona pellucida*, is a glyco-protein shell, which captures and tethers the sperm before they penetrate it. . . . The sperm is captured at the initial contact between the sperm tip and the *zona*. . . . Since the thrust [of the sperm] is much smaller than the force needed to break a single affinity bond, the first bond made upon the tip-first meeting of the sperm and *zona* can result in the capture of the sperm.[39]

Experiments in another lab reveal similar patterns of data interpretation. Gerald Schatten and Helen Schatten set out to show that, contrary to conventional wisdom, the 'egg is not merely a large, yolk-filled sphere into which the sperm burrows to endow new life. Rather, recent research suggests the almost heretical view that sperm and egg are mutually active partners'.[40] This sounds like a departure from the stereotypical textbook view, but further reading reveals Schatten and Schatten's conformity to the aggressive-sperm metaphor. They describe how 'the sperm and egg first touch when, from the tip of the sperm's triangular head, a long, thin filament shoots out and harpoons the egg'. Then we learn that 'remarkably, the harpoon is not so much fired as assembled at great speed, molecule by molecule, from a pool of protein stored in a specialized region called the acrosome. The filament may grow as much as twenty times longer than the sperm head itself before its tip reaches the egg and sticks.'[41] why not call this 'making a bridge' or 'throwing out a line' rather than firing a harpoon? Harpoons pierce prey and injure or kill them, while this filament only sticks. And why not focus, as the Hopkins lab did, on the stickiness of the egg, rather than the stickiness of the sperm? Later in the article, the Schattens replicate the common view of the sperm's perilous journey into the warm darkness of the vagina, this time for the purpose of explaining its journey into the egg itself: '[The sperm] still has an arduous journey ahead. It must penetrate farther into the egg's huge sphere of cytoplasm and somehow locate the nucleus, so that the two cells' chromosomes can fuse. The sperm dives down into the cytoplasm, its tail beating. But it is soon interrupted by the sudden and swift migration of the egg nucleus, which rushes toward the sperm with a velocity triple that of the movement of chromosomes during cell division, crossing the entire egg in about a minute.[42]

Like Schatten and Schatten and the biophysicists at Johns Hopkins, another researcher has recently made discoveries that seem to point to a more interactive view of the relationship of egg and sperm. This work, which Paul Wassarman conducted on the sperm and eggs of mice, focuses on identifying the specific molecules in the egg coat (the zona pellucida) that are involved in egg–sperm interaction. At first glance, his descriptions seem to fit the model of an egalitarian relationship. Male and female gametes 'recognize one another', and 'interactions . . . take place between sperm and egg'.[43] But the article in *Scientific American* in which those descriptions appear begins with a vignette that presages the dominant motif of their presentation: 'It has been more than a century since Hermann Fol, a Swiss zoologist, peered into his microscope and became the first person to see a sperm penetrate an egg, fertilize it and form the first cell of a new embryo'.[44] This portrayal of the sperm as the active party—the one that *penetrates* and *fertilizes* the egg and *produces* the embryo—is not cited as an example of an earlier, now outmoded view. In fact, the author reiterates the point later in the article: 'Many sperm can bind to and penetrate the zona pellucida, or outer coat, of an unfertilized mouse egg, but only one sperm will eventually fuse with the thin plasma membrane surrounding the egg proper (*inner sphere*), fertilizing the egg and giving rise to a new embryo'.[45]

The imagery of sperm as aggressor is particularly startling in this case: the main discovery being reported is isolation of a particular molecule *on the egg coat* that plays an important role in fertilization! Wassarman's choice of language sustains the picture. He calls the molecule that has been isolated, ZP3, a 'sperm receptor'. By allocating the passive, waiting role to the egg, Wassarman can continue to describe the sperm as the actor, the one that makes it all happen: 'The basic process begins when many sperm first attach loosely and then bind tenaciously to receptors on the surface of the egg's thick outer coat, the zona pellucida. Each sperm, which has a large number of egg-binding proteins on its surface, binds to many sperm receptors on the egg. More specifically, a site on each of the egg-binding proteins fits a complementary site on a sperm receptor, much as a key fits a lock.'[46] With the sperm designated as the 'key' and the egg the 'lock', it is obvious which one acts and which one is acted upon. Could this imagery not be reversed, letting the sperm (the lock) wait until the egg produces the key? Or could we speak of two halves of a locket matching, and regard the matching itself as the action that initiates the fertilization?

It is as if Wassarman were determined to make the egg the receiving partner. Usually in biological research, the *protein* member of the pair of binding molecules is called the receptor, and physically it has a pocket in it rather like a lock. As the diagrams that illustrate Wassarman's article show, the molecules on the sperm are proteins and have 'pockets'. The small, mobile molecules that fit into these pockets are called ligands. As shown in the diagrams, ZP3 on the egg is a polymer of 'keys'; many small knobs stick out. Typically, molecules on the sperm would be called receptors and molecules on the egg would be called ligands. But Wassarman chose to name ZP3 on the egg the receptor and to create a new term, 'the egg-binding protein', for the molecule on the sperm that otherwise would have been called the receptor.[47]

. . .

SOCIAL IMPLICATIONS: THINKING BEYOND

All three of these revisionist accounts of egg and sperm cannot seem to escape the hierarchical imagery of older accounts. Even though each new account gives the egg a larger and more active role, taken together they bring into play another cultural stereotype: woman as a dangerous and aggressive threat. In the Johns Hopkins lab's revised model, the egg ends up as the female aggressor who 'captures and tethers' the sperm with her sticky zona, rather like a spider lying in wait in her web.[48] The Schatten lab has the egg's nucleus 'interrupt' the sperm's dive with a 'sudden and swift' rush by which she 'clasps the sperm and guides its nucleus to the center'.[49] Wassarman's description of the surface of the egg 'covered with thousands of plasma membrane-bound projections, called microvilli' that reach out and clasp the sperm adds to the spiderlike imagery.[50]

These images grant the egg an active role but at the cost of appearing disturbingly aggressive. Images of woman as dangerous and aggressive, the *femme fatale* who victimizes men, are widespread in Western literature and culture.[51] More specific is the connection of spider imagery with the idea of an engulfing, devouring mother.[52] New data did not lead scientists to eliminate gender stereotypes in their descriptions of egg and sperm. Instead, scientists simply began to describe egg and sperm in different, but no less damaging, terms.

Can we envision a less stereotypical view? Biology itself provides another model that could be applied to the egg and the sperm. The cybernetic model—with its feedback loops, flexible adaptation to change, co-ordination of the parts within a whole, evolution over time, and changing response to the environment—is common in genetics, endocrinology, and ecology and has a growing influence in medicine in general.[53] This model has the potential to shift our imagery from the negative, in which the female reproductive system is castigated both for not producing eggs after birth and for producing (and thus wasting) too many eggs overall, to something more positive. The female reproductive system could be seen as responding to the environment (pregnancy or menopause), adjusting to monthly changes (menstruation), and flexibly changing from reproductivity after puberty to non-reproductivity later in life. The sperm and egg's interaction could also be described in cybernetic terms. J. F. Hartman's research in reproductive biology demonstrated fifteen years ago that if an egg is killed by being pricked with a needle, live sperm cannot get through the zona.[54] Clearly, this evidence shows that the egg and sperm *do* interact on more mutual terms, making biology's refusal to portray them that way all the more disturbing.

We would do well to be aware, however, that cybernetic imagery is hardly neutral. In the past, cybernetic models have played an important part in the imposition of social control. These models inherently provide a way of thinking about a 'field' of interacting components. Once the field can be seen, it can become the object of new forms of knowledge, which in turn can allow new forms of social control to be exerted over the components of the field. During the 1950s, for example, medicine began to recognize the psychosocial *environment* of the patient: the patient's family and its psychodynamics. Professions such as social work began to focus on this new environment, and the resulting knowledge became one way to further control the patient. Patients began to be seen not as isolated, individual bodies, but as psychosocial entities located in an 'ecological' system: management of 'the patient's psychology was a new entrée to patient control'.[55]

The models that biologists use to describe their data can have important social effects. During the nineteenth century, the social and natural sciences strongly influenced each other: the social ideas of Malthus about how to avoid the natural increase of the poor inspired Darwin's *Origin of Species*.[56] Once the *Origin* stood as a description of the natural world, complete with competition and market struggles, it could be reimported into social science as social Darwinism, in order to justify the social order of the time. What we are seeing now is similar: the importation of cultural ideas about passive females and heroic males into the 'personalities' of gametes. This amounts to the 'implanting of social imagery on representations of nature so as to lay a firm basis for reimporting exactly that same imagery as natural explanations of social phenomena'.[57]

Further research would show us exactly what social effects are being wrought from the biological imagery of egg and sperm. At the very least, the imagery keeps alive some of the hoariest old stereotypes about weak damsels in distress and their strong male rescuers. That these stereotypes are now being written in at the level of the *cell* constitutes a powerful move to make them seem so natural as to be beyond alteration.

The stereotypical imagery might also encourage people to imagine that what results from the interaction of egg and sperm— a fertilized egg—is the result of deliberate 'human' action at the cellular level. Whatever the intentions of the human couple, in this microscopic 'culture' a cellular 'bride' (or *femme fatale*) and a cellular 'groom' (her victim) make a cellular baby. Rosalind Petchesky points out that through visual representations such as sonograms, we are given '*images* of younger and younger, and tinier and tinier, fetuses being "saved" '. This leads to 'the point of visibility being "pushed back" *indefinitely*'.[58] Endowing egg and sperm with intentional action, a key aspect of personhood in our culture, lays the foundation for the point of viability being pushed back to the moment of fertilization. This will likely lead to greater acceptance of technological developments and new forms of scrutiny and manipulation, for the benefit of these inner 'persons': court-ordered restrictions on a pregnant woman's activities in order to protect her fetus, fetal surgery, amniocentesis, and rescinding of abortion rights, to name but a few examples.[59]

Even if we succeed in substituting more egalitarian, interactive metaphors to describe the activities of egg and sperm, and manage to avoid the pitfalls of cybernetic models, we would still be guilty of endowing cellular entities with personhood. More crucial, then, than what *kinds* of personalities we bestow on cells is the very fact that we are doing it at all. This process could ultimately have the most disturbing social consequences.

One clear feminist challenge is to wake up sleeping metaphors in science, particularly those involved in descriptions of the egg and the sperm. Although the literary convention is to call such metaphors 'dead', they are not so much dead as sleeping, hidden within the scientific content of texts—and all the more powerful for it.[60] Waking up such metaphors, by becoming aware of when we are projecting cultural imagery onto what we study, will improve our ability to investigate and understand nature. Waking up such metaphors, by becoming aware of their implications, will rob them of their power to naturalize our social conventions about gender.

Notes

1. James Hillman, *The Myth of Analysis* (Evanston, Ill.: Northwestern University Press, 1972), 220.
2. The textbooks I consulted are the main ones used in classes for undergraduate premedical students or medical students (or those held on reserve in the library for these classes) during the past few years at Johns Hopkins University. These texts are widely used at other universities in the country as well.
3. Arthur C. Guyton, *Physiology of the Human Body*, 6th edn. (Philadelphia: Saunders College Publishing, 1984), 624.
4. Arthur J. Vander, James H. Sherman, and Dorothy S. Luciano, *Human Physiology: The Mechanisms of Body Function*, 3d edn. (New York: McGraw-Hill, 1980), 483–4.
5. Vernon B. Mountcastle, *Medical Physiology*, 14th edn. (London: Mosby, 1980), 2:1624.
6. Eldra Pearl Solomon, *Human Anatomy and Physiology* (New York: CBS College Publishing, 1983), 678.
7. For elaboration, see Emily Martin, *The Woman in the Body: A Cultural Analysis of Reproduction* (Boston: Beacon, 1987), 27–53.
8. Vander, Sherman, and Luciano, 568.
9. Melvin Konner, 'Childbearing and Age', *New York Times Magazine* (27 Dec. 1987), 22–3, esp. 22.
10. In her essay 'Have Only Men Evolved?' (in Sandra Harding and Merrill B.

Hintikka (eds.), *Discovering Reality: Feminist Perspectives on Epistemology, Metaphysics, Methodology, and Philosophy of Science* (Dordrecht: Reidel, 1983), 45–69, esp. 60–1), Ruth Hubbard points out that sociobiologists have said the female invests more energy than the male in the production of her large gametes, claiming that this explains why the female provides parental care. Hubbard questions whether it 'really takes more "energy" to generate the one or relatively few eggs than the large excess of sperms required to achieve fertilization'.

11. See Carol Delaney, 'The Meaning of Paternity and the Virgin Birth Debate', *Man*, 21/3 (Sept. 1986), 494–513. She discusses the difference between this scientific view that women contribute genetic material to the fetus and the claim of long-standing Western folk theories that the origin and identity of the fetus comes from the male, as in the metaphor of planting a seed in soil.

12. For a suggested direct link between human behaviour and purportedly passive eggs and active sperm, see Erik H. Erikson, 'Inner and Outer Space: Reflections on Womanhood', *Daedalus*, 93/2 (Spring 1964), 582–606, esp. 591.

13. Guyton (n. 3 above), 619; and Mountcastle (n. 5 above), 1609.

14. Jonathan Miller and David Pelham, *The Facts of Life* (New York: Viking Penguin, 1984), 5.

15. Bruce Alberts *et al.*, *Molecular Biology of the Cell* (New York: Garland, 1983), 796.

16. Ibid. 796.

17. See, e.g. William F. Ganong, *Review of Medical Physiology*, 7th edn. (Los Altos, Calif.: Lange Medical Publications, 1975), 322.

18. Alberts *et al.* (n. 15 above), 796.

19. Guyton, 615.

20. Solomon (n. 6 above), 683.

21. Vander, Sherman, and Luciano (n. 4 above), 4th edn. (1985), 580.

22. Alberts *et al.*, 796.

23. All biology texts quoted above use the word 'penetrate'.

24. Solomon, 700.

25. A. Deldecus *et al.*, 'The Importance of Feminist Critique for Contemporary Cell Biology', *Hypatia*, 3/1 (Spring 1988), 61–76.

26. Gerald Schatten and Helen Schatten, 'The Energetic Egg', *Medical World News*, 23 (23 Jan. 1984), 51–3, esp. 51.

27. Alberts *et al.*, 796.

28. Guyton (n. 3 above), 613.

29. Miller and Pelham (n. 14 above), 7.

30. Alberts *et al.* (n. 15 above), 804.

31. Ibid. 801.

32. Ruth Herschberger, *Adam's Rib* (New York: Pelligrini & Cudaby, 1948), esp. 84.

33. Bennett M. Shapiro, 'The Existential Decision of a Sperm', *Cell*, 49/3 (May 1987), 293–4, esp. 293.

34. Ludwik Fleck, *Genesis and Development of a Scientific Fact*, ed. Thaddeus J. Trenn and Robert K. Merton (Chicago: University of Chicago Press, 1979), 38.

35. Jay M. Baltz carried out the research I describe when he was a graduate student in the Thomas C. Jenkins Department of Biophysics at Johns Hopkins University.

36. Far less is known about the physiology of sperm than comparable female

substances, which some feminists claim is no accident. Greater scientific scrutiny of female reproduction has long enabled the burden of birth control to be placed on women. In this case, the researchers' discovery did not depend on development of any new technology. The experiments made use of glass pipettes, a manometer, and a simple microscope, all of which have been available for more than one hundred years.

37. Jay Baltz and Richard A. Cone, 'What Force Is Needed to Tether a Sperm?' (abstract for Society for the Study of Reproduction, 1985), and 'Flagellar Torque on the Head Determines the Force Needed to Tether a Sperm' (abstract for Biophysical Society, 1986).
38. Jay M. Baltz, David F. Katz, and Richard A. Cone, 'The Mechanics of the Sperm–Egg Interaction at the Zona Pellucida', *Biophysical Journal*, 54/4 (Oct. 1988), 643–54.
39. Ibid. 643, 650.
40. Schatten and Schatten (n. 26 above), 51.
41. Ibid. 52.
42. Schatten and Schatten, 53.
43. Paul M. Wassarman, 'Fertilization in Mammals', *Scientific American*, 259/6 (Dec. 1988), 78–84, esp. 78, 84.
44. Ibid. 78.
45. Ibid. 79.
46. Ibid. 78.
47. Since receptor molecules are relatively *immotile* and the ligands that bind to them relatively *motile*, one might imagine the egg being called the receptor and the sperm the ligand. But the molecules in question on egg and sperm are immotile molecules. It is the sperm as a *cell* that has motility, and the egg as a cell that has relative immotility.
48. Baltz, Katz, and Cone (n. 38 above), 643, 650.
49. Schatten and Schatten, 53.
50. Paul M. Wassarman, 'The Biology and Chemistry of Fertilization', *Science*, 235/4788 (30 Jan. 1987), 553–60, esp. 554.
51. Mary Ellman, *Thinking about Women* (New York: Harcourt Brace Jovanovich, 1968), 140; Nina Auerbach, *Woman and the Demon* (Cambridge, Mass.: Harvard University Press, 1982), esp. 186.
52. Kenneth Alan Adams, 'Arachnophobia: Love American Style', *Journal of Psychoanalytic Anthropology*, 4/2 (1981), 157–97.
53. William Ray Arney and Bernard Bergen, *Medicine and the Management of Living* (Chicago: University of Chicago Press, 1984).
54. J. F. Hartman, R. B. Gwatkin, and C. F. Hutchison, 'Early Contact Interactions between Mammalian Gametes *In Vitro*', *Proceedings of the National Academy of Sciences (US)* 69/10 (1972), 2767–9.
55. Arney and Bergen, 68.
56. Ruth Hubbard, 'Have Only Men Evolved?' (n. 10 above), 51–2.
57. David Harvey, personal communication, Nov. 1989.
58. Rosalind Petchesky, 'Fetal Images: The Power of Visual Culture in the Politics of Reproduction', *Feminist Studies*, 13/2 (Summer 1987), 263–92, esp. 272.
59. Rita Arditti, Renate Klein, and Shelley Minden, *Test-Tube Women* (London: Pandora, 1984); Ellen Goodman, 'Whose Right to Life?' *Baltimore Sun* (17 Nov. 1987); Tamar Lewin, 'Courts Acting to Force Care of the Unborn', *New York Times* (23 Nov. 1987), A1 and B10; Susan Irwin and Brigitte Jordan,

Knowledge, Practice, and Power: Court Ordered Cesarean Sections', *Medical Anthropology Quarterly*, 1/3 (Sept. 1987), 319–34.
60. Thanks to Elizabeth Fee and David Spain, who in February 1989 and April 1989, respectively, made points related to this.

Part III. Language, Gender, and Science

8 Race and Gender: The Role of Analogy in Science

Nancy Leys Stepan

Metaphor occupies a central place in literary theory, but the role of metaphors, and of the analogies they mediate, in scientific theory is still debated.[1] One reason for the controversy over metaphor, analogy, and models in science is the intellectually privileged status that science has traditionally enjoyed as the repository of nonmetaphorical, empirical, politically neutral, universal knowledge. During the scientific revolution of the seventeenth century, metaphor became associated with the imagination, poetic fancy, subjective figures, and even untruthfulness and was contrasted with truthful, unadorned, objective knowledge—that is, with science itself.[2]

...

One result of the dichotomy established between science and metaphor was that obviously metaphoric or analogical science could only be treated as 'prescientific' or 'pseudoscientific' and therefore dismissable.[3] Because science has been identified with truthfulness and empirical reality, the metaphorical nature of much modern science tended to go unrecognized. And because it went unrecognized, as Colin Turbayne has pointed out, it has been easy to mistake the model in science 'for the thing modelled'—to think, to take his example, that nature *was* mechanical, rather than to think it was, metaphorically, seen as mechanical.[4]

...

Although the role of metaphor and analogy in science is now recognized, a critical theory of scientific metaphor is only just being elaborated. The purpose of this article is to contribute to the development of such a theory by using a particular analogy in the history of the life sciences to explore a series of related questions concerning the cultural sources of scientific analogies, their role in scientific reasoning, their normative consequences, and the process by which they change.

Reprinted with permission from *ISIS*, 77 (1986), 261–77.

RACE AND GENDER: A POWERFUL SCIENTIFIC ANALOGY

The analogy examined is the one linking race to gender, an analogy that occupied a strategic place in scientific theorizing about human variation in the nineteenth and twentieth centuries.

As has been well documented, from the late Enlightenment on students of human variation singled out racial differences as crucial aspects of reality, and an extensive discourse on racial inequality began to be elaborated.[5] In the nineteenth century, as attention turned increasingly to sexual and gender differences as well, gender was found to be remarkably analogous to race, such that the scientist could use racial difference to explain gender difference, and vice versa.[6]

Thus it was claimed that women's low brain weights and deficient brain structures were analogous to those of lower races, and their inferior intellectualities explained on this basis.[7] Woman, it was observed, shared with Negroes a narrow, childlike, and delicate skull, so different from the more robust and rounded heads characteristic of males of 'superior' races. Similarly, women of higher races tended to have slightly protruding jaws, analogous to, if not as exaggerated as, the apelike, jutting jaws of lower races.[8] Women and lower races were called innately impulsive, emotional, imitative rather than original, and incapable of the abstract reasoning found in white men.[9] Evolutionary biology provided yet further analogies. Woman was in evolutionary terms the 'conservative element' to the man's 'progressive', preserving the more 'primitive' traits found in lower races, while the males of higher races led the way in new biological and cultural directions.[10]

Thus when Carl Vogt, one of the leading German students of race in the middle of the nineteenth century, claimed that the female skull approached in many respects that of the infant and in still further respects that of lower races, whereas the mature male of many lower races resembled in his 'pendulous' belly a Caucasian woman who had had many children, and in his thin calves and flat tights the ape, he was merely stating what had become almost a cliché of the science of human difference.[11]

So fundamental was the analogy between race and gender that the major modes of interpretation of racial traits were invariably evoked to explain sexual traits. For instance, just as scientists spoke of races as distinct 'species', incapable of crossing to produce viable 'hybrids', scientists analysing male–female differences sometimes spoke of females as forming a distinct 'species', individual members of which were in danger of degenerating into psychosexual hybrids when they tried to cross the boundaries proper to their sex.[12] Darwin's theory of sexual selection was applied to both racial and sexual difference, as was the neo-Lamarckian theory of the American Edward D. Cope.[13] A last, confirmatory example of the analogous place of gender and race in scientific theorizing is taken from the history of hormone biology. Early in the twentieth century the anatomist and student of race Sir Arthur Keith interpreted racial differences in the human species as a function of pathological disturbances of the newly discovered 'internal secretions' or hormones. At about the same time, the apostle of sexual frankness and well-known student of sexual variation Havelock Ellis used internal secretions to explain the small, but to him vital, differences in the physical and psychosexual make-up of men and women.[14]

In short, lower races represented the 'female' type of the human species, and females the 'lower race' of gender. As the example from Vogt indicates, however, the analogies concerned more than race and gender. Through an intertwined and overlapping series of analogies, involving often quite complex comparisons, identifications, cross-references, and evoked associations, a variety of 'differences'—physical and psychical, class and national—were brought together in a biosocial science of human variation. By analogy with the so-called lower races, women, the sexually deviate, the criminal, the urban poor, and the insane were in one way or another constructed as biological races apart whose differences from the white male, and likenesses to each other, 'explained' their different and lower position in the social hierarchy.[15]

It is not the aim of this article to provide a systematic history of the biosocial science of racial and sexual difference based on analogy. The aim is rather to use the race–gender analogy to analyse the nature of analogical reasoning in science itself. When and how did the analogy appear in science? From what did it derive its scientific authority? How did the analogy shape research? What did it mean when a scientist claimed that the mature male of many lower races resembled a mature Caucasian female who had had many children? No simple theory of resemblance or substitution explains such an analogy. How did the analogy help construct the very similarities and differences supposedly 'discovered' by scientists in nature? What theories of analogy and metaphor can be most effectively applied in the critical study of science?

THE CULTURAL SOURCES OF SCIENTIFIC METAPHOR

. . .

The origin of many of the 'root metaphors' of human difference are obscure. G. Lakoff and M. Johnson suggest that the basic values of a culture are usually compatible with 'the metaphorical structure of the most fundamental concepts in the culture'.[16] Not surprisingly, the social groups represented metaphorically as 'other' and 'inferior' in Western culture were socially 'disenfranchised' in a variety of ways, the causes of their disenfranchisement varying from group to group and from period to period. Already in ancient Greece, Aristotle likened women to the slave on the grounds of their 'natural' inferiority. Winthrop Jordan has shown that by the early Middle Ages a binary opposition between blackness and whiteness was well established in which blackness was identified with baseness, sin, the devil, and ugliness, and whiteness with virtue, purity, holiness, and beauty.[17] Over time, black people themselves were compared to apes, and their childishness, savageness, bestiality, sexuality, and lack of intellectual capacity stressed. The 'Ethiopian', the 'African', and especially the 'Hottentot' were made to stand for all that the white male was not; they provided a rich analogical source for the understanding and representation of other 'inferiorities'. In his study of the representation of insanity in Western culture, for instance, Gilman shows how the metaphor of blackness could be borrowed to explicate the madman, and vice versa. In similar analogical fashion, the labouring poor were represented as the 'savages' of Europe, and the criminal as a 'Negro'.

When scientists in the nineteenth century, then, proposed an analogy between racial and sexual differences, or between racial and class differences, and began to generate new data on the basis of such analogies, their interpretations of human difference and similarity were widely accepted, partly because of their fundamental congruence with cultural expectations. In this particular science, the metaphors and analogies were not strikingly new but old, if unexamined and diffuse. The scientists' contribution was to elevate hitherto unconsciously held analogies into self-conscious theory, to extend the meanings attached to the analogies, to expand their range via new observations and comparisons, and to give them precision through specialized vocabularies and new technologies. Another result was that the analogies became 'naturalized' in the language of science, and their metaphorical nature disguised.

In the scientific elaboration of these familiar analogies, the study of race led the way, in part because the differences between blacks and whites seemed so 'obvious', in part because the abolition movement gave political urgency to the issue of racial difference and social inequality. From the study of race came the association between inferiority and the ape. The facial angle, a measure of hierarchy in nature obtained by comparing the protrusion of the jaws in apes and man, was widely used in analogical science once it was shown that by this measure Negroes appeared to be closer to apes than the white race.[18] Established as signs of inferiority, the facial angle and blackness could then be extended analogically to explain other inferior groups and races. For instance, Francis Galton, Darwin's cousin and the founder of eugenics and statistics in Britain, used the Negro and the apish jaw to explicate the Irish: 'Visitors to Ireland after the potato famine', he commented, 'generally remarked that the Irish type of face seemed to have become more prognathous, that is, more like the negro in the protrusion of the lower jaw.'[19]

Especially significant for the analogical science of human difference and similarity were the systematic study and measurement of the human skull. The importance of the skull to students of human difference lay in the fact that it housed the brain, differences in whose shape and size were presumed to correlate with equally presumed differences in intelligence and social behaviour. It was measurements of the skull, brain weights, and brain convolutions that gave apparent precision to the analogies between anthropoid apes, lower races, women, criminal types, lower classes, and the child. It was race scientists who provided the new technologies of measurement—the callipers, cephalometers, craniometers, craniophores, craniostats, and parietal goniometers.[20] The low facial angles attributed by scientists starting in the 1840s and 1850s to women, criminals, idiots, and the degenerate, and the corresponding low brain weights, protruding jaws, and incompletely developed frontal centres where the higher intellectual faculties were presumed to be located, were all taken from racial science. By 1870 Paul Topinard, the leading French anthropologist after the death of Paul Broca, could call on data on sexual and racial variations from literally hundreds of skulls and brains, collected by numerous scientists over decades, in order to draw the conclusion that Caucasian women were indeed more prognathous or apelike in their jaws than white men, and even the largest women's brains, from the 'English or Scotch' race, made them like the African male.[21] Once 'woman' had been shown to be indeed analogous to lower races by the new science of anthropometry and had become, in essence, a racialized category, the traits and qualities special to woman could in turn be used in an analogical understanding of lower races. The analogies now had the weight of empirical reality and scientific theory. The similarities between a Negro and a white woman, or between a criminal and a Negro, were realities of nature, somehow 'in' the individuals studied.

METAPHORIC INTERACTIONS

We have seen that metaphors and analogies played an important part in the science of human difference in the nineteenth century. The question is, what part? I want to suggest that the metaphors functioned as the science itself—that without them the science did not exist. In short, metaphors and analogies can be constituent elements of science.

It is here that I would like to introduce, as some other historians of science have done, Max Black's 'interaction' theory of metaphor, because it seems that the metaphors discussed in this essay, and the analogies they mediated, functioned like interaction metaphors, and that thinking about them in these terms clarifies their role in science.[22]

By interaction metaphors, Black means metaphors that join together and bring into cognitive and emotional relation with each other two different things, or systems of things, not normally so joined. Black follows I. A. Richards in opposing the 'substitution' theory of metaphor, in which it is supposed that the metaphor is telling us indirectly something factual about the two subjects—that the metaphor is a *literal comparison*, or is capable of a literal translation in prose. Richards proposed instead that 'when we use a metaphor, we have two thoughts of different things active together and supported by a single word or phrase, whose meaning is the resultant of their interaction.' Applying the interaction theory to the metaphor 'The poor are the negroes of Europe', Black paraphrases Richards to claim that 'our thoughts about the European poor and American negroes are "active together" and "interact" to produce a meaning that is a resultant of that interaction.'[23] In such

a view, the metaphor cannot be simply reduced to literal comparisons or 'like' statements without loss of meaning or cognitive content, because meaning is a product of the interaction between the
two parts of a metaphor.
. . .

Black's point is that by their interactions and evoked associations
both parts of a metaphor are changed. Each part is seen as more like
the other in some characteristic way. Black was primarily interested
in ordinary metaphors of a culture and in their commonplace associations. But instead of commonplace associations, a metaphor
may evoke more specially constructed systems of implications.
Scientists are in the business of constructing exactly such systems of
implications, through their empirical investigations into nature
and through their introduction into discourse of specialized vocabularies and technologies.[24] It may be, indeed, that what makes an
analogy suitable for scientific purposes is its ability to be suggestive
of new systems of implications, new hypotheses, and therefore new
observations.[25]
In the case of the nineteenth-century analogical science of
human difference, for instance, the system of implications evoked
by the analogy linking lower races and women was not just a gener-
alized one concerning social inferiority, but the more precise and
specialized one developed by years of anthropometric, medical, and
biological research. When 'woman' and 'lower races' were analogically and routinely joined in the anthropological, biological, and
medical literature of the 1860s and 1870s, the metaphoric interactions involved a complex system of implications about similarity
and difference, often involving highly technical language (for example, in one set of measurements of the body in different races cited
by Paul Topinard in 1878 the comparisons included measures in
each race of their height from the ground to the acromion, the epi-
condyle, the styloid process of the radius, the great trochanter, and
the internal malleolus). The systems of implications evoked by the
analogy included questions of comparative health and disease
(blacks and women were believed to show greater degrees of insanity and neurasthenia than white men, especially under conditions
of freedom), of sexual behaviour (females of 'lower races' and
lower-class women of 'higher races', especially prostitutes, were
believed to show similar kinds of bestiality and sexual promiscuity,
as well as similar signs of pathology and degeneracy such as
deformed skulls and teeth), and of 'childish' characteristics, both
physical and moral.[26]

As already noted, one of the most important systems of implications about human groups developed by scientists in the nineteenth century on the basis of analogical reasoning concerned head shapes and brain sizes. It was assumed that blacks, women, the lower classes, and criminals shared low brain weights or skull capacities. Paul Broca, the founder of the Société d'Anthropologie de Paris in 1859, asserted: 'In general, the brain is larger in mature adults than in the elderly, in men than in women, in eminent men than in men in mediocre talent, in superior races than in inferior races. . . . Other things being equal, there is a remarkable relationship between the development of intelligence and the volume of the brain.'[27]

Such a specialized system of implications based on the similarities between brains and skulls appeared for the first time in the phrenological literature of the 1830s. Although analogies between women and blackness had been drawn before, woman's place in nature and her bio-psychological differences from men had been discussed by scientists mainly in terms of reproductive function and sexuality, and the most important analogies concerned black females (the 'sign' of sexuality) and lower-class or 'degenerate' white women. Since males of all races had no wombs, no systematic, apparently scientifically validated grounds of comparison between males of 'lower' races and women of 'higher' races existed.

Starting in the 1820s, however, the phrenologists began to focus on differences in the shape of the skull of individuals and groups, in the belief that the skull was a sign faithfully reflecting the various organs of mind housed in the brain, and that differences in brain organs explained differences in human behaviour. And it is in the phrenological literature, for almost the first time, that we find women and lower races compared directly on the basis of their skull formations. In their 'organology', the phrenologists paid special attention to the organ of 'philoprogenitiveness', or the faculty causing 'love of offspring', which was believed to be more highly developed in women than men, as was apparent from their more highly developed upper part of the occiput. The same prominence, according to Franz Joseph Gall, was found in monkeys and was particularly well developed, he believed, in male and female Negroes.[28]

By the 1840s and 1850s the science of phrenology was on the wane, since the organs of the brain claimed by the phrenologists did not seem to correspond with the details of brain anatomy as described by neurophysiologists. But although the specific conclusions of the phrenologists concerning the anatomical structure and functions of the brain were rejected, the principle that differences in individual and group function were products of differences in the shape and size of the head was not. This principle underlay the claim that some measure, whether of cranial capacity, the facial angle, the brain volume, or brain weight, would be found that would provide a true indicator of innate capacity, and that by such a measure women and lower races would be shown to occupy analogous places in the scale of nature (the 'scale' itself of course being a metaphorical construct).

By the 1850s the measurement of women's skulls was becoming an established part of craniometry and the science of gender joined analogically to race. Vogt's *Lectures on Man* included a long discussion of the various measures available of the skulls of men and women of different races. His data showed that women's smaller brains were analogous to the brains of lower races, the small size explaining both groups' intellectual inferiority. (Vogt also concluded that within Europe the intelligentsia and upper classes had the largest heads, and peasants the smallest.)[29] Broca shared Vogt's interest; he too believed it was the smaller brains of women and 'lower' races, compared with men of 'higher' races, that caused their lesser intellectual capacity and therefore their social inferiority.[30]

One novel conclusion to result from scientists' investigations into the different skull capacities of males and females of different races was that the gap in head size between men and women had apparently widened over historic time, being largest in the 'civilized' races such as the European, and smallest in the most savage races.[31] The growing difference between the sexes from the prehistoric period to the present was attributed to evolutionary, selective pressures, which were believed to be greater in the white races than the dark and greater in men than women. Paradoxically, therefore, the civilized European woman was less like the civilized European man than the savage man was like the savage woman. The 'discovery' that the male and female bodies and brains in the lower races were very alike allowed scientists to draw direct comparisons between a black male and white female. The male could be taken as representative of both sexes of his race and the black female could be virtually ignored in the analogical science of intelligence, if not sexuality.

Because interactive metaphors bring together a *system* of implications, other features previously associated with only one subject in the metaphor are brought to bear on the other. As the analogy between women and race gained ground in science, therefore, women were found to share other points of similarity with lower races. A good example is prognathism. Prognathism was a measure of the protrusion of the jaw and of inferiority. As women and lower races became analogically joined, data on the 'prognathism' of females were collected and women of 'advanced' races implicated in this sign of inferiority. Havelock Ellis, for instance, in the late nineteenth-century bible of male–female differences *Man and Woman*, mentioned the European woman's slightly protruding jaw as a trait, not of high evolution, but of the lower races, although he added that in white women the trait, unlike in the lower races, was 'distinctly charming'.[32]

Another set of implications brought to bear on women by analogy with lower races concerned dolichocephaly and brachycephaly, or longheadedness and roundheadedness. Africans were on the whole more longheaded than Europeans and so dolichocephaly was generally interpreted as signifying inferiority. Ellis not surprisingly found that on the whole women, criminals, the degenerate, the insane, and prehistoric races tended to share with dark races the more narrow, dolichocephalic heads representing an earlier (and by implication, more primitive) stage of brain development.[33]

ANALOGY AND THE CREATION OF NEW KNOWLEDGE

In the metaphors and analogies joining women and the lower races, the scientist was led to 'see' points of similarity that before had gone unnoticed. Women became more 'like' Negroes, as the statistics on brain weights and body shapes showed. The question is, what kind of 'likeness' was involved?

Here again the interaction theory of metaphor is illuminating. As Black says, the notion of similarity is ambiguous. Or as Stanley Fish puts it, 'Similarity is not something one finds but something one must establish.'[34] Metaphors are not meant to be taken literally but they do imply some structural similarity between the two things joined by the metaphor, a similarity that may be new to the readers of the metaphoric or analogical text, but that they are culturally capable of grasping.

However, there is nothing obviously similar about a white woman of England and an African man, or between a 'criminal type' and a 'savage'. (If it seems to us as though there is, that is because the metaphor has become so woven into our cultural and linguistic system as to have lost its obviously metaphorical quality and to seem a part of 'nature'.) Rather it is the metaphor that permits us to see similarities that the metaphor itself helps constitute.[35] The metaphor, Black suggests, 'selects, emphasizes, suppresses and organizes features' of reality, thereby allowing us to see new connections between the two subjects of the metaphor, to pay attention to details hitherto unnoticed, to emphasize aspects of human experience otherwise treated as unimportant, to make new features into 'signs' signifying inferiority.[36] It was the metaphor joining lower races and women, for instance, that gave significance to the supposed differences between the shape of women's jaws and those of men.

. . .

The metaphor, in short, served as a programme of research. Here the analogy comes close to the idea of a scientific 'paradigm' as elaborated by Kuhn in *The Structure of Scientific Revolutions*; indeed Kuhn himself sometimes writes of paradigms as though they are extended metaphors and has proposed that 'the same interactive, similarity-creating process which Black has isolated in the functioning of metaphor is vital also in the function of models in science.'[37]

The ability of an analogy in science to create new kinds of knowledge is seen clearly in the way the analogy organizes the scientists' understanding of causality. Hesse suggests that a scientific metaphor, by joining two distinct subjects, implies more than mere structural likeness. In the case of the science of human difference, the analogies implied a similar *cause* of the similarities between races and women and of the differences between both groups and white males. To the phrenologists, the cause of the large organs of philoprogenitiveness in monkeys, Negroes, and women was an innate brain structure. To the evolutionists, sexual and racial differences were the product of slow, adaptive changes involving variation and selection, the results being the smaller brains and lower capacities of the lower races and women, and the higher intelligence and evolutionarily advanced traits in the males of higher races. Barry Barnes suggests we call the kind of 'redescription' involved in a metaphor or analogy of the kind being discussed here an 'explanation', because it forces the reader to 'understand' one aspect of reality in terms of another.[38]

NANCY LEYS STEPAN

ANALOGY AND THE SUPPRESSION OF KNOWLEDGE

Especially important to the functioning of interactive metaphors in science is their ability to neglect or even suppress information about human experience of the world that does not fit the similarity implied by the metaphor. In their 'similarity-creating' capacity, metaphors involve the scientist in a selection of those aspects of reality that are compatible with the metaphor. This selection process is often quite unconscious. Stephen Jay Gould is especially telling about the ways in which anatomists and anthropologists unselfconsciously searched for and selected measures that would prove the desired scales of human superiority and inferiority and how the difficulties in achieving the desired results were surmounted.

Gould has subjected Paul Broca's work on human differences to particularly thorough scrutiny because Broca was highly regarded in scientific circles and was exemplary in the accuracy of his measurements. Gould shows that it is not Broca's measurements *per se* that can be faulted, but rather the ways in which he unconsciously manipulated them to produce the very similarities already 'contained' in the analogical science of human variation. To arrive at the conclusion of women's inferiority in brain weights, for example, meant failing to make any correction for women's smaller body weights, even though other scientists of the period were well aware that women's smaller brain weights were at least in part a function of their smaller body sizes. Broca was also able to 'save' the scale of ability based on head size by leaving out some awkward cases of large-brained but savage heads from his calculations, and by somehow accounting for the occasional small-brained 'geniuses' from higher races in his collection.[39]
. . .

When contrary evidence could not be ignored, it was often reinterpreted to express the fundamental valuations implicit in the metaphor. Gould provides us with the example of neoteny, or the retention in the adult of childish features such as a small face and hairlessness. A central feature of the analogical science of inferiority was that adult women and lower races were more childlike in their bodies and minds than white males. But Gould shows that by the early twentieth century it was realized that neoteny was a positive feature of the evolutionary process. 'At least one scientist,

132

Havelock Ellis, did bow to the clear implication and admit the superiority of women, even though he wriggled out of a similar confession for blacks.' As late as the 1920s the Dutch scientist Louis Bolk, on the other hand, managed to save the basic valuation of white equals superior, blacks and women equal inferior by 'rethinking' the data and discovering after all that blacks departed more than whites from the most favourable traits of childhood.[40]

To reiterate, because a metaphor or analogy does not directly present a pre-existing nature but instead helps 'construct' that nature, the metaphor generates data that conform to it, and accommodates data that are in apparent contradiction to it, so that nature is seen via the metaphor and the metaphor becomes part of the logic of science itself.[41]

. . .

A BRIEF CONCLUSION

In this essay I have indicated only some of the issues raised by a historical consideration of a specific metaphoric or analogical science. There is no attempt at completeness or theoretical closure. My intention has been to draw attention to the ways in which metaphor and analogy can play a role in science, and to show how a particular set of metaphors and analogies shaped the scientific study of human variation. I have also tried to indicate some of the historical reasons why scientific texts have been 'read' non-metaphorically, and what some of the scientific and social consequences of this have been.

Some may argue I have begged the question of metaphor and analogy in science by treating an analogical science that was 'obviously pseudoscientific'. I maintain that it was not obviously pseudoscientific to its practitioners, and that they were far from being at the periphery of the biological and human sciences in the nineteenth and early twentieth centuries. I believe other studies will show that what was true for the analogical science of human difference may well be true also for other metaphors and analogies in science.

My intention has also been to suggest that a theory of metaphor is as critical to science as it is to the humanities. We need a critical theory of metaphor in science in order to expose the metaphors by which we learn to view the world scientifically, not because these metaphors are necessarily 'wrong', but because they are so powerful.

Notes

1. A metaphor is a figure of speech in which a name or descriptive term is transferred to some object that is different from, but analogous to, that to which it is properly applicable. According to Max Black, 'every metaphor may be said to mediate an analogy or structural correspondence': see Black, 'More About Metaphor', in Andrew Ortony (ed.), *Metaphor and Thought* (Cambridge: Cambridge University Press, 1979), 19–43, on p. 31. In this article, I have used the terms *metaphor* and *analogy* interchangeably.
2. G. Lakoff and M. Johnson, *Metaphors We Live By* (Chicago/London: University of Chicago Press, 1980), 191. Scientists' attacks on metaphor as extrinsic and harmful to science predate the Scientific Revolution.
3. For this point see Jamie Kassler, 'Music as a Model in Early Science', *History of Science*, 20 (1982), 103–39.
4. Colin M. Turbayne, *The Myth of Metaphor* (Columbia: University of South Carolina Press, 1970), 24.
5. See Nancy Stepan, *The Idea of Race in Science: Great Britain, 1800–1960* (London: Macmillan, 1982), esp. ch. 1.
6. No systematic history of the race–gender analogy exists. The analogy has been remarked on, and many examples from the anthropometric, medical, and embryological sciences provided, in Stephen Jay Gould, *The Mismeasure of Man* (New York: W. W. Norton, 1981), and in John S. Haller and Robin S. Haller, *The Physician and Sexuality in Victorian America* (Urbana: University of Illinois Press, 1974).
7. Haller and Haller, *The Physician and Sexuality*, 48–9, 54. Among the several craniometric articles cited by the Hallers, see esp. J. McGrigor Allan, 'On the Real Differences in the Minds of Men and Women', *Journal of the Anthropological Society of London*, 7 (1869), cxcv–ccviii, on p. cciv; and John Cleland, 'An Inquiry into the Variations of the Human Skull', *Philosophical Transactions, Royal Society*, 89 (1870), 117–74.
8. Havelock Ellis, *Man and Woman: A Study of Secondary Sexual Characters* (London: A. & C. Black, 6th edn. 1926), 106–7.
9. Herbert Spencer, 'The Comparative Psychology of Man', *Popular Science Monthly*, 8 (1875–6), 257–69.
10. Ellis, *Man and Woman* (cit. n. 8), 491.
11. Carl Vogt, *Lectures on Man: His Place in Creation, and in the History of the Earth* (London: Longman, Green, & Roberts, 1864), 81.
12. James Weir, 'The Effect of Female Suffrage on Posterity', *American Naturalist*, 29 (1895), 198–215.
13. Charles Darwin, *The Descent of Man, and Selection in Relation to Sex* (London: John Murray, 1871), ii, chs. 17–20; Edward C. Cope, 'The Developmental Significance of Human Physiognomy', *American Naturalist*, 17 (1883), 618–27.
14. Arthur Keith, 'Presidential Address: On Certain Factors in the Evolution of Human Races', *Journal of the Royal Anthropological Institute*, 64 (1916), 10–33; Ellis, *Man and Woman* (cit. n. 8), p. xii.

15. See Nancy Stepan, 'Biological Degeneration: Races and Proper Places', in J. Edward Chamberlin and Sander L. Gilman (eds.), *Degeneration: The Dark Side of Progress* (New York: Columbia University Press, 1985), 97–120, esp. 112–13. For an extended exploration of how various stereotypes of difference intertwined with each other, see Sander L. Gilman, *Difference and Pathology: Stereotypes of Sexuality, Race, and Madness* (Ithaca, NY: Cornell University Press, 1985).

16. Lakoff and Johnson, *Metaphors We Live By* (cit. n. 2), 22. The idea of root metaphors is Stephen Pepper's in *World Hypothesis* (Berkeley/Los Angeles: University of California Press, 1966), 91.

17. Winthrop D. Jordan, *White over Black: American Attitudes toward the Negro, 1550–1812* (New York: Norton, 1977), 7.

18. Stepan, *The Idea of Race in Science*, 6–10.

19. Francis Galton, 'Hereditary Improvement', *Fraser's Magazine*, 7 (1873), 116–30.

20. These instruments and measurements are described in detail in Paul Topinard, *Anthropology* (London: Chapman & Hall, 1878), pt. 2, chs. 1–4.

21. Ibid. 311.

22. Max Black, *Models and Metaphor* (Ithaca, NY: Cornell University Press, 1961), esp. chs. 3 and 13. See also Mary Hesse, *Models and Analogies in Science* (Notre Dame, Ind.: University of Notre Dame Press, 1966); Mary Hesse, 'The Explanatory Function of Metaphor', in Y. Bar-Hillel (ed.), *Logic, Methodology and Philosophy of Science* (Amsterdam: North-Holland, 1965), 249–59; and Richard Boyd, 'Metaphor and Theory Change: What is "Metaphor" a Metaphor For?', in A. Ortony (ed.), *Metaphor and Thought*. 356–408.

23. Black, *Models and Metaphor*, 38, quoting I. A. Richards, *Philosophy of Rhetoric* (Oxford: Oxford University Press, 1938), 93.

24. See Turbayne, *Myth of Metaphor* (cit. n. 4), p. 19, on this point.

25. Black himself believed scientific metaphors belonged to the pretheoretical stage of a discipline. Here I have followed Boyd, who argues in 'Metaphor and Theory Change' (cit. n. 22), p. 357, that metaphors can play a role in the development of theories in relatively mature sciences. Some philosophers would reserve the term 'model' for extended, systematic metaphors in science.

26. For an example of the analogous diseases and sexuality of 'lower' races and 'lower' women, see Eugene S. Talbot, *Degeneracy: Its Causes, Signs, and Results* (London: Walter Cott, 1898), 18, 319–23.

27. Paul Broca, 'Sur le volume et la forme du cerveau suivant les individus et suivant les races', *Bulletin de la Société d'Anthropologie Paris*, 2 (1861), 304.

28. Franz Joseph Gall, 'The Propensity to Philoprogenitiveness', *Phrenological Journal*, 2 (1824–5), 20–33.

29. Vogt, *Lectures on Man* (cit. n. 11), 88. Vogt was quoting Broca's data.

30. Gould, *Mismeasure of Man* (cit. n. 6), 103.

31. Broca's work on the cranial capacities of skulls taken from three cemeteries in Paris was the most important source for this conclusion. See his 'Sur la capacité des crânes parisiens des divers époques', *Bull. Soc. Anthr. Paris*, 3 (1862), 102–16.

32. Ellis, *Man and Woman* (cit. n. 8), 106–7.

33. Alexander Sutherland, 'Woman's Brain', *Nineteenth Century*, 47 (1900), 802–10; and Ellis, *Man and Woman*, 98. Ellis was on the whole, however,

cautious about the conclusions that could be drawn from skull capacities and brain weights.

34. Stanley Fish, 'Working on the Chain Gang: Interpretation in the Law and Literary Criticism' in W. J. T. Mitchell (ed.), *The Politics of Interpretation* (Chicago: University of Chicago Press, 1983), 277.

35. Max Black, as cited in Ortony, *Metaphor and Thought* (cit. n. 1), 5.

36. Black, *Models and Metaphor* (cit. n. 7), 44.

37. Thomas S. Kuhn, *The Structure of Scientific Revolutions* (Chicago: University of Chicago Press, 2nd edn., 1973) esp. ch. 4; and Thomas S. Kuhn, 'Metaphor in Science', in A. Ortony (ed.) *Metaphor and Thought*, 409–19, on 415.

38. Barry Barnes, *Scientific Knowledge and Sociological Theory* (London: Routledge & Kegan Paul, 1974), 49.

39. Gould, *Mismeasure of Man* (cit. n. 6), 73–112. For another example see Stephen Jay Gould, 'Morton's Ranking of Race by Cranial Capacity', *Science*, 200 (1978), 503–509.

40. Gould, *Mismeasure of Man*, 120–1.

41. Terence Hawkes, *Metaphor* (London: Methuen, 1972), 88, suggests that metaphors 'will retrench or corroborate as much as they expand our vision', thus stressing the normative, consensus-building aspects of metaphor.

9

Why Mammals are Called Mammals: Gender Politics in Eighteenth-Century Natural History

Londa Schiebinger

In 1758, in the tenth edition of his *Systema naturae*, Carolus Linnaeus introduced the term *Mammalia* into zoological taxonomy. For his revolutionary classification of the animal kingdom—hailed in the twentieth century as the starting point of modern zoological nomenclature—Linnaeus devised this word, meaning literally 'of the breast', to distinguish the class of animals embracing humans, apes, ungulates, sloths, sea-cows, elephants, bats, and all other organisms with hair, three ear bones, and a four-chambered heart.[1] In so doing, he made the female mammae the icon of that class.
...

It is possible, however, to see the Linnaean coinage as a political act. The presence of milk-producing mammae is, after all, but one characteristic of mammals, as was commonly known to eighteenth-century European naturalists. Furthermore, the mammae are 'functional' in only half of this group of animals (the females) and, among those, for a relatively short period of time (during lactation) or not at all. Linnaeus could have derived a term from a number of equally unique, and perhaps more universal, characteristics of the class he designated mammals, choosing *Pilosa* (the hairy ones—although the significance given hair, and especially beards, was also saturated with gender), for example, or *Aurecaviga* (the hollow-eared ones).
...

It has been said that God created nature and Linnaeus gave it order.[2] Carolus Linnaeus, also known as Carl von Linné, 'Knight of the Order of the Polar Star',[3] was the central figure in developing European taxonomy and nomenclature. His *Systema naturae* treated the three classical kingdoms of nature—animal, vegetable, and mineral—growing from a folio of only twelve pages in 1735 to

This essay is part of my book *Nature's Body: Gender in the Making of Modern Science* (Boston, 1993) and is reprinted with permission.

a three-volume work of 2400 pages in the twelfth and last edition revised by Linnaeus himself in 1766. In the epoch-making tenth edition, Linnaeus gave binomial names (generic and specific) to all the animals known to him, nearly 4400 species.[4]

Linnaeus divided animals into six classes: *Mammalia, Aves, Amphibia, Pisces, Insecta,* and *Vermes.*[5] Although Linnaeus had based important aspects of plant taxonomy on sexual dimorphism, the class *Mammalia* was the only one of his major zoological divisions to focus on reproductive organs and the only term to highlight a characteristic associated primarily with the female. The names of his other classes came, in many cases, from Aristotle: *Aves* simply means bird; *Amphibia* emphasizes habitat; *Insecta* refers to the segmentation of the body. *Vermes* derives from the colour (red-brown) of the common earthworm. Scientific nomenclature was a conservative enterprise in the eighteenth century; suitable terms tended to be conserved and new terms derived by modifying traditional ones. Linnaeus, however, broke with tradition by creating the term *Mammalia.*

In coining the term mammals, Linnaeus abandoned Aristotle's canonical term, *Quadrupedia.* For more than two thousand years, most of the animals we now designate as mammals (along with most reptiles and several amphibians) had been called quadrupeds.

. . .

Aristotelian categories and terminology remained fundamental to European natural history well into the early modern period.

. . .

Natural historians before Linnaeus had struggled long and hard with the problems of classification. John Ray, often credited with developing binomial nomenclature (although he did not employ it systematically), had used the term *Vivipara* to unite whales and other aquatic mammals with terrestrial quadrupeds. Within his subcategory *Terrestria,* he suggested the term *Pilosa* (hairy animals) as more comprehensive than *Quadrupedia* and thus more suitable for joining amphibious manatees with land-dwelling quadrupeds.[6] Peter Artedi, Linnaeus's close friend and colleague, had called attention to hair in his proposed *Trichozoologia,* or 'science of the hirsute animal'.[7] Linnaeus might well have chosen the more traditional adjective *Pilosa* for his new class of quadrupeds; in his system, hair had the same diagnostic value as mammae.[8] All mammals (including the whale) have hair, and it is still today considered a distinguishing characteristic of mammals.

But Linnaeus did not draw on tradition; he devised instead a new term, *Mammalia*. In its defence, Linnaeus remarked that even if his critics did not believe that humans originally walked on all fours, surely every man born of woman must admit that he was nourished by his mother's milk.[9] Linnaeus thus called attention to the fact, commonly known since Aristotle, that hairy, viviparous females lactate. Linnaeus was also convinced of the diagnostic value of the teat. As early as 1732, in his *Tour of Lapland*, he had already announced, 'If I knew how many teeth and of what peculiar form each animal has, as well as how many udders and where situated, I should perhaps be able to contrive a most natural methodical arrangement of quadrupeds.'[10] In the first edition of his *Systema naturae*, he used the number and position of teats or udders to align orders within his class of *Anthropomorpha* (complicating factors being that females and males often have different numbers and that females of the same species may also vary in the number of their teats).[11] In 1758, Linnaeus announced the term *Mammalia* in the tenth edition of his *Systema naturae* with the words, 'Mammalia, these and no other animals have mammae [mammata].' He seemed quite unconcerned that mammae were not a universal characteristic of the class he intended to distinguish. 'All females', he wrote on the following page, 'have lactiferous mammae of determinate number, as do males (except for the horse).'[12]

Mammalia resonated with the older term *animalia*, derived from *anima*, meaning the breath of life or vital spirit.[13] The new term also conformed to Linnaeus's own rules for zoological terms: it was pleasing to the ear, easy to say and to remember, and not more than twelve letters long.[14] For the rest of his life, Linnaeus fiddled with his system, moving animals from order to order, creating new categories and combinations to better capture nature's order. Yet he never rechristened mammals.

The term *Mammalia* gained almost immediate acceptance.[15] There were, however, detractors of note. Buffon scorned the entire project of taxonomy but especially Linnaean taxonomy and nomenclature. For Buffon, the task of the natural historian was to describe each animal precisely—its mode of reproduction, nourishment, customs, and habitat—not to divide nature's bounty into artificial groups with incomprehensible names of Greek or Latin origin. Buffon took particular offence at the prominence Linnaeus gave the breast: 'A general character, such as the teat, taken to identify quadrupeds should at least belong to all quadrupeds.' (Buffon, like Linnaeus, recognized that stallions, for example, have no teats.)[16] Buffon also complained that Linnaeus's order *Anthropomorpha* lumped together things as different as humans, apes, and sloths. This 'violence' was wreaked on the natural scheme of things, he lamented, all because there was 'some small relationship between the number of nipples or teeth of these animals or some slight resemblance in the form of their horns'.[17]

Other taxonomists, including Felix Vicq-d'Azyr and Thomas Pennant, continued to use the traditional term, *Quadrupedia*. Still others developed their own alternatives. The Frenchman Henri de Blainville in 1816 tried to rationalize zoological nomenclature, renaming mammals *Pilifera* (having hair), birds *Pennifera* (having feathers), and reptiles *Squammifera* (having scales).[18] In England, John Hunter proposed the term *Tetracoilia*, drawing attention to the four-chambered heart.[19]

These critics met with little success. . . . Linnaeus's term *Mammalia* was retained even after the Darwinian revolution and is today recognized by the International Code of Zoological Nomenclature.

The word 'mamma'—the singular form of 'mammae', designating the milk-secreting organs of the female—probably derives from baby talk, being a reduplicated syllable often uttered by young children, who in many countries are taught to use it as their word for mother.[20] Linnaeus devised the term *Mammalia* from the Latin *mammae*, intending it to refer to the breast or teat itself as much as to its milk-producing aspects. These terms—breast and teat—are somewhat ambiguous. Teat sometimes refers to the nipple of a cow, sheep, or goat but also refers to the internal structures of the mammary gland. In humans (and some birds), breast refers to the chest area as well as to the milk-producing organ in the female. Today, it is the mammary gland with its milk-producing structures that defines the class *Mammalia*. Two groups fit uncomfortably in this taxon: males, with their dry and barren vestigial breasts, and monotremes (egg-laying mammals such as the duckbilled platypus, spiny echidna, and anteater), which have mammary glands but no nipples.[21]

The question of why males have breasts at all has long plagued naturalists. The eighteenth-century medical doctor Louis de Jaucourt addressed this issue as one of six basic questions about the breast in his article, 'Mamelle', for Diderot and d'Alembert's *Encyclopédie*. Jaucourt, who also wrote a well-known entry on 'Femme', noted that the particular cast of the human body and its parts answered to nature's need to conserve the species and that even though some parts, such as male breasts, may be superfluous, nature did not take them away. He was quick to argue that male breasts are not defective, that in many cases milk flows in great abundance from them. That males rarely produce milk was to be traced to the absence of menstrual blood—the source of milk. According to Jaucourt, with the onset of puberty, blood surges throughout the female body, causing young women's breasts to 'inflate'; the passion of love also experienced at this age causes them to inflate even further. Men do not have menses, the author continued, and therefore their breasts—though anatomically similar to women's—never inflate.[22]

The fanciful notion that males are, indeed, capable of producing milk was popular among naturalists. Aristotle had considered it an omen of extraordinary good fortune when a male goat produced milk in such quantities that cheese could be made from it.[23] Eighteenth-century naturalists reported the secretion of a fatty milky substance—'witch's milk'—from the breasts of male as well as female newborns. Buffon related many examples of the male breast filling with milk at the onset of puberty. A boy of fifteen, for example, pressed from one of his breasts more than a spoonful of 'true' milk.[24] John Hunter offered the example of a father who nursed his eight children. This man began nursing when his wife was unable to satisfy a set of twins. 'To soothe the cries of the male child', Hunter wrote, 'the father applied his left nipple to the infant's mouth, who drew milk from it in such quantity as to be nursed in perfectly good health.' (The father also shared with his wife all other domestic duties.) Considering milk production within the bounds of normal male physiology, Hunter dutifully noted that the man 'was not a hermaphrodite'.[25]

Despite dramatic examples such as these, most naturalists recognized that the male breast was barren. Why, then, did males have breasts at all? Erasmus Darwin, Charles Darwin's grandfather, suggested that the vestigial male teat lent credence to Plato's theory that mammals had hermaphroditic origins and only later developed into distinct males and females.[26] Late into the nineteenth century, comparative anatomists continued to embrace the notion that some remote progenitor of the vertebrate kingdom had been androgynous.[27] Charles Darwin, following Clémence Royer, suggested that in an earlier age male mammals had aided females in nursing their offspring and that later, some pattern of events (such as smaller litters) rendered male assistance unnecessary. The disuse of the organ led to its becoming vestigial, and this was passed on to future generations.[28] Today, naturalists emphasize that many organs in the female and male, such as the clitoris and penis, and the labia majora and scrotal sac, are identical in the early embryos and only later—after the action of various hormones—develop along different paths.[29]

. . .

Were there good reasons for Linnaeus to name mammals mammals? This question implies a logic uncharacteristic of the naming process. Names of taxa collect over time, and unless there is a technical problem—as was the case with the term *Quadrupedia*—they pass unchanged from generation to generation. Naturalists also name plants and animals for other than empirical reasons. Plants or animals that are pleasing are often named after a wife or colleague, while a particularly odious species might be given the name of a professional rival (for instance, *Siegesbeckia*, a small and unpleasant flowering weed that Linnaeus named after Johann Siegesbeck, a critic of his sexual system).[30]

Zoological nomenclature—like all language—is to some degree arbitrary; naturalists devise convenient terms to identify groups of animals. But nomenclature is also historical, growing out of specific contexts, conflicts, and circumstances. The historian can fairly ask why a certain term was coined. In coining the term *Mammalia*, Linnaeus intended to highlight an essential trait of that class of animals. Geoffroy Saint-Hilaire and Georges Cuvier, in their article 'Mammalogie' for the *Magazin encyclopédique* of 1795, summed up the practice of eighteenth-century taxonomists, stating that primary organs determine classes, while secondary organs determine orders. In 1827, Cuvier continued to argue that the mammae distinguish the class bearing their name better than any other external characteristic.[31]

Is Cuvier's statement, in fact, true? Does the longevity of Linnaeus's term reflect the fact that he was simply right, that the mammae do represent a primary, universal, and unique characteristic of mammals (as would have been a parlance of the eighteenth century)? Yes and no. Paleontologists today identify the mammary gland as one of at least six uniquely mammalian characteristics.[32] Still, Linnaeus was perhaps overly exuberant in singling out the breast or teat itself—a sexually charged part of the female body— rather than its function. One could argue that the term *Lactantia* (the lactating ones, derived from Linnaeus's own description of female mammae) would have better captured the significance of the mammae; certainly, Linnaeus was wrong to think that the number and position of the teats themselves were significant. But *Lactantia* still refers exclusively to females. *Lactentia* or *Sugentia* (both meaning 'the sucking ones') would have better universalized the term, since male as well as female young suckle at their mothers' breasts.

The fact remains that the mammae were only one among several traits that could have been highlighted. Even by eighteenth-century criteria, there was not one characteristic alone that could determine class assignment. As Buffon recognized, species—defined for sexually reproducing organisms as members of a group of individuals that can produce fertile offspring—is the only taxon that exists in nature.[33] This does not mean that higher units—genera, families, orders, classes, and on up—are arbitrary; these must be consistent with evolutionary genealogy.[34] Yet, as we have seen, Linnaeus could have chosen from equally valid terms such as *Pilosa, Aurecaviga, Lactentia,* or *Sugentia.* Because Linnaeus had choices, I suggest that his focus on the breast responded to broader cultural and political trends.

Long before Linnaeus, the female breast had been a powerful icon in Western cultures, representing both the sublime and bestial in human nature.[35] The grotesque, withered breasts on witches and devils represented temptations of wanton lust, sins of the flesh, and humanity fallen from paradise. The firm spherical breasts of Aphrodite, the Greek ideal, represented an otherworldly beauty and virginity. During the French Revolution, the bared female breast—embodied in the strident Marianne—became a resilient symbol of freedom.[36] From the multi breasted Diana of Ephesus to the fecund-bosomed Nature, the breast symbolized generation, regeneration, and renewal.

Linnaeus created his term *Mammalia* in response to the question of humans' place in nature. In his quest to find an appropriate term for (what we would call) a taxon uniting humans and beasts, Linnaeus made the breast—and specifically the fully developed female breast—the icon of the highest class of animals. It might be argued that, by privileging a uniquely female characteristic in this way, Linnaeus broke with long-standing traditions that saw the male as the measure of all things. In the Aristotelian tradition, the female had been seen as a misbegotten male, a monster or error of nature. By honouring the mammae as sign and symbol of the highest class of animals, Linnaeus assigned a new value to the female, especially women's unique role in reproduction.

It is important to note, however, that in the same volume in which Linnaeus introduced the term *Mammalia*, he also introduced the name *Homo sapiens*. This term, 'man of wisdom', was used to distinguish humans from other primates (apes, lemurs, and bats, for example). In the language of taxonomy, *sapiens* is what is known as a 'trivial' name. (Linnaeus at one point pondered the choice of the name *Homo diurnus*, designed to contrast with *Homo nocturnus*.) From a historical point of view, however, the choice of the term *sapiens* is highly significant. 'Man' had traditionally been distinguished from animals by his reason; the medieval apposition, *animal rationale*, proclaimed his uniqueness.[37] Thus, within Linnaean terminology, a female characteristic (the lactating mamma) ties humans to brutes, while a traditionally male characteristic (reason) marks our separateness.

The notion that woman—lacking male perfections of mind and body—resides nearer the beast is an ancient one. Among all the organs of a woman's body, her reproductive organs were considered most animal-like. For Plato, the uterus was an animal with its own sense of smell, wandering within the female body and leaving disease and destruction in its path.[38] Galen and even Vesalius (for a time) reported that the uterus had horns. The milk production of the female breast had already been taken to link humans with animals. Aristotle, in his *Historia animalium*, had recognized that all internally viviparous animals—women, sheep, horses, cows, and whales, for example—nurse their young.

. . .

Myths and legends also portrayed suckling as a point of close connection between humans and beasts, suggesting the interchangeability of human and animal breasts in this respect. A nanny-goat, Amaltheia, was said to have nursed the young Zeus.[39] A she-wolf served as the legendary nurse to Romulus and Remus, the founders of Rome. From the Middle Ages to the seventeenth and eighteenth centuries, bears and wolves were reported to have suckled abandoned children. Children were thought to imbibe certain characteristics of the animals that nursed them—the 'wild Peter' found in northern Germany in 1724 grew thick hair all over his body as a result of his nurturance at the breast of a bear. Linnaeus believed that ancient heroes, put to the breast of the lioness, absorbed her great courage along with her milk.[40]

In rarer instances, humans were reported even to have suckled animals. Veronica Giuliani, beatified by Pius II (1405–64), took a real lamb to bed with her and suckled it at her breast in memory of the lamb of God.[41] European voyagers reported that native South

Americans kept their breasts active by letting animals of all kinds feed from them. In Siam, women were said to have suckled apes.[42] The practice of animals suckling at human breasts was also reported in Europe. William Godwin recorded that as Mary Wollstonecraft lay dying after childbirth, the doctor forbade the child the breast and 'procurred puppies to draw off the milk'.[43]

Linnaeus thus followed well-established Western conceptions when he suggested that women belong to nature in ways that men do not.[44] As Carolyn Merchant has shown, nature itself has been conceived as female in most Western intellectual traditions.[45] The identification of woman with the fecund and nurturing qualities of nature was highlighted in the influential eighteenth-century artists and engravers Hubert François Gravelot and Charles Cochin's personification of Nature as a virgin, her breasts dripping with milk.[46]

It is significant that Linnaeus used the mammiferous Diana of the Ephesians, an ancient symbol of animal and human fertility, as the frontispiece to his *Fauna Svecica*, where he first defended his inclusion of humans among the quadrupeds. Linnaeus's Diana, half captive in the fecund earth, emerges to display her womb—the centre of life—and her nourishing breasts.[47] In this classic image, her curiously immobilized trunk is covered with symbols of both fertility (bees, acorns, bulls, crabs) and chastity (stags, lions, roses). Her pendulous breasts, heavy with milk, represent the life-force of nature, mother and nurse of all living things.

For Linnaeus to suggest, then, that humans shared with animals the capacity to suckle their young was nothing new. This uniquely female feature had long been considered less than human. But it had also been considered more than human. In the Christian world, milk had been seen as providing sustenance—for both body and spirit. Throughout the Middle Ages, the faithful cherished vials of the Virgin's milk as a healing balm, a symbol of mercy, an eternal mystery. As Marina Warner has pointed out, the Virgin Mary endured none of the bodily pleasures and pains associated with childbearing (menstruation, sexual intercourse, pregnancy, or labour) except for suckling. The tender Madonna suckled the infant Jesus both as his historical mother and as the metaphysical image of the nourishing Mother Church.[48] During the twelfth century, maternal imagery—especially suckling and nurturing—extended also to church fathers. Abbots and prelates were encouraged to 'mother' the souls in their charge, to expose their breasts and let their bosoms expand with the milk of consolation.[49] Even the full breasts of God the Father were said to be milked by the Holy Spirit into the cup of the Son of God.[50]

In subcurrents of religious thought, mother's milk was thought to impart knowledge. Philosophia-Sapientia, the personification of wisdom, suckled philosophers at her breasts moist with the milk of knowledge and moral virtue. Augustine of Hippo, too, imagined himself drinking from the breasts of Sapientia. Centuries later, men of science still sought the secrets of (female) nature within her bosom, though with a rather different purpose. Goethe waxed poetic on the point: 'Infinite Nature, where are thy breasts, those well-springs of all life on which hang heaven and earth, toward which my withered breast strains?'[51] For Goethe, at least, the scientist's new desire was not to suckle at the breast of nature but to imitate its nourishing power.

. . .

In a certain sense, Linnaeus's focus on the milk-bearing breast was at odds with trends that found beauty (though not necessarily salvation) in the virginal breast. In both Greek and Christian traditions, the ideal breast was an unused one—small, firm, and spherical; the process of milk swelling the breast was thought to deform it. Mythical female figures—the goddesses Artemis and Aphrodite, the martial Amazons (who supposedly burned away one breast so that their bows would lie flat against their chests), and the nursing mother of Christ—were all virgins. Of all the female Virtues, only Charity possessed a non-virginal body: infants drank maternal bounty, love, and humility from her breasts.[52]

The classic aesthetic ideal of the firm, unused breast was realized in the bodies of many upper-class medieval and early modern European women who avoided the burden of suckling their own children. . . . Wealthy women in Europe bore children but most often did not nurse them. For this task, women were employed who were considered closer to nature: peasants and, in overseas colonies, native women and women of African descent ('often but one remove above a brute', in the words of one observer).[53] Even when, late in the eighteenth century, fashionable women did for a while nurse their infants, the shape and size of the breast was at issue. Moderately sized, nicely oval breasts with small but protuberant nipples were thought to produce better milk than large, pendulous breasts.[54]

. . .

Europeans' fascination with the female breast provided a receptive climate for Linnaeus's new term. But more immediate political concerns compelled him to focus scientific attention on the mammae.

His scientific vision arose alongside important political trends in the eighteenth century—the restructuring of both child care and women's lives as mothers, wives, and citizens. The stress he placed on the naturalness of a mother giving suck to her young reinforced the social movements undermining the public power of women and attaching a new value to mothering. Despite the Enlightenment credo that all 'men' were by nature equal, middle-class women were not to become fully enfranchised citizens or professionals in the state but newly empowered mothers within the home.

Most directly, Linnaeus joined the campaign to abolish the ancient custom of wet nursing, . . . preparing a dissertation against the evils of wet nursing in 1752 just a few years before coining the term *Mammalia*. . . . His work titled 'Step Nurse' (translated into French as 'La nourrice marâtre, ou dissertation sur les suites funestes du nourrissage mercénaire') sounded the themes of the Enlightenment attack on wet nursing.[55] First and foremost, wet nursing violated the laws of nature. Nature—herself 'a tender and provident mother'—had set the course for female reproduction; digression from her laws endangered both mother and child.

. . .

In this pamphlet, Linnaeus also foreshadowed his subsequent nomenclature by contrasting the barbarity of women who deprived their children of mother's milk with the gentle care of great beasts—the whale, the fearsome lioness, and fierce tigress—who willingly offer their young the teat.[56]

. . .

For the enlightened savant, the laws of nature dictated more than the rules for reproductive regimes: they also dictated social order. Medical authority, the legal system, and popular literature worked together to create new interest in maternal breast-feeding. As pre-scribed in Rousseau's influential novel *Emile*, breast-feeding became fashionable among French upper-class women for a short period in the late eighteenth century.[57] In France and Germany, leading medical doctors advocated laws that would force healthy women to nurse their own infants.[58] The French National Convention decreed in 1793 that only mothers who nursed their own children would be eligible for state aid (women in poor health were exempted). Similar laws were put into effect in Prussia in 1794, just a few years after Frederick the Great installed a modern version of Diana of the Ephesians in his Potsdam garden.[59]

Authors of anti-wet-nursing literature . . . were highly moralistic about returning women to what was considered their rightful place as loving and caring mothers. . . . It is remarkable that in the heady days of the French Revolution, when revolutionaries marched behind the martial and bare-breasted Liberty,[60] the maternal breast became nature's sign that women belonged only in the home. Delegates to the French National Convention used the breast as a natural sign that women should be barred from citizenship and the wielding of public power. In this case, 'the breasted ones' were to be confined to the home. In denying women political power, Pierre-Gaspard Chaumette, an official of the Paris Commune, asked indignantly: 'Since when is it permitted to abandon one's sex? Since when is it decent for women to forsake the pious care of their households and the cribs of their children, coming instead to public places, to hear speeches in the galleries and senate? Is it to men that nature confided domestic cares? Has she given us breasts to feed our children?'[61]

This message was embodied in the 'Festival of Unity and Indivisibility' of 1793, celebrating the first anniversary of the Republic. Jacques-Louis David's carefully orchestrated festival featured a 'Fountain of Regeneration' built on the ruins of the Bastille, the symbol of absolutism. As described in the popular press, eighty-six (male) deputies to the National Convention drank joyfully from the spouting breasts of 'Nature' personified as Isis, the Egyptian goddess of fertility. While the male deputies publicly drank the maternal 'milk' of national renewal from the breasts of the colossal Isis, exemplary republican mothers quietly re-enacted the scene, giving their virtuous milk to future citizens of the state.

The year 1793 marked the fateful repression of women's demands for active citizenship and also, as Lynn Hunt has shown, a turning point in republican images of women. When publicly represented, women were no longer cast as the strident Marianne, the symbol of Liberty, but increasingly in motherly roles. Festivals featured parades of pregnant women; women in ceremonies, such as the Festival of the Supreme Being of 1794, were all wives and mothers, many pressing nurslings to their breasts.[62]

Linnaeus's term *Mammalia* helped legitimize the sexual division of labour in European society by emphasizing how natural it was for females—both human and non-human—to suckle and rear their own offspring. Linnaean systematics had sought to render nature universally comprehensible, yet the categories he devised infused nature with middle-class European notions of gender. Linnaeus saw the females of all species as tender mothers, a vision he (wittingly or unwittingly) projected onto Europeans' understandings of nature. This was not the only instance in which Linnaeus suffused nature with parochial notions of gender. In his botanical taxonomy—for which he was hailed the father of modern botany—Linnaeus established (hetero)sexuality as the key to classification. In so doing, as I have shown elsewhere, he gave male parts priority over female parts in determining the status of an organism in the plant kingdom, imposing traditional notions of gender hierarchy onto science.[63]

In naming mammals, there is no evidence that Linnaeus intentionally chose a gender-charged term; he may have done so naïvely. But he did not do so arbitrarily. The fact that scientists might be innocent of the implications of their work does not make them any less mediators or marketeers of political ideas. Historians have to appreciate the contingency of scientific knowledge and especially what is foregone in the choice of one particular course over another. This is why the political historian of science asks: Why do we know this and not that? Who gains from knowledge of this and not that?

The story of the origins of the term *Mammalia* provides yet another example of how science is not value neutral but emerges from complex cultural matrices. The term Linnaeus coined in 1758 solved the problem of how to classify the whale with its terrestrial congeners and did away with Aristotle's outmoded term quadruped. But, more than that, it provided a solution to the place of humankind within nature and ultimately of womankind within European culture.

Notes

1. The 10th edition of Linnaeus's *Systema naturae* (1758) and Carl Clerck's *Aranei Svecici* (1757) together form the starting point of modern zoological nomenclature. See W. D. L. Ride (ed.), *International Code of Zoological Nomenclature* (London, 1985), 1: 3. The term *Mammalia* first appeared in a student dissertation, *Natura pelagi*, in 1757 but was not published until 1760. *Amoenitates academicae* (Erlangen, 1788), 5: 68–77.
2. Albrecht von Haller rather mockingly called him 'the second Adam'. Gunnar Broberg, 'Linnaeus and Genesis,' in Broberg, *Linnaeus: Progress and Prospects in Linnaean Research* (Stockholm, 1980), 34.
3. Marie-Jean Caritat, Marquis de Condorcet, 'Eloge de M. de Linné, *Histoire de l'Académie Royale des Sciences* (Paris, 1778), 66. Linnaeus was the first man of letters to be awarded this honour.
4. W. T. Stearn, 'The Background of Linnaeus's Contributions to the Nomenclature and Methods of Systematic Biology', *Systematic Zoology*, 8 (1959), 4–22; and E. G. Linsley and R. L. Usinger, 'Linnaeus and the Development of the International Code of Zoological Nomenclature', *ibid.* 39–46.

5. Carl Linnaeus, *Systema naturae per regna tria naturae*, 10th edn. (Stockholm, 1758).

6. John Ray, *Synopsis methodica animalium quadrupedum et serpentini generis*, 'Animalium tabula generalis' (London, 1693), 53. See also William Gregory, 'Linnaeus as an Intermediary between Ancient and Modern Zoology', *Annals of the New York Academy of Sciences*, 18 (1908), 21–31, esp. 25. Ray's terms were used as adjectives, not nouns—an important distinction at a time when scholastics still distinguished between essence and accident. Theodor Gill, 'The Story of a Word—Mammal', *Popular Science Monthly*, 61 (1902), 434–8.

7. Gunnar Broberg, '*Homo sapiens*: Linnaeus's Classification of Man', in Tore Frangsmyr (ed.), *Linnaeus: The Man and His Work* (Berkeley, Calif., 1983), 175.

8. I have derived this term from Linnaeus's use of *pilus* in his catalogue of mammalian traits (*Systema naturae*, 10th edn., 12).

9. Gunnar Broberg, *Homo sapiens L.: Studier i Carl von Linnes naturuppfattning och människolära* (Stockholm, 1975), 176.

10. Carl Linnaeus, *Lachesis Lapponica; or, A Tour in Lapland*, trans. James E. Smith (London, 1811), i. 191, slightly modified.

11. Pig nipples, for example, vary from between eight and eighteen in number. Ernst Bresslau, *The Mammary Apparatus of the Mammalia in the Light of Ontogenesis and Phylogenesis* (London, 1920), 98.

12. Linnaeus, *Systema naturae* (1758), 14, 16.

13. Gill, 'Story of a Word—Mammal', 435.

14. Stearn, 'Background of Linnaeus's Contributions to the Nomenclature', 8.

15. Linnaeus's term *Primates* encountered more resistance. Notably, Johann Friedrich Blumenbach and Georges Cuvier insisted on separating humans and apes into distinct orders. Blumenbach coined the term *Inermis* (without weapons) for humans, and Cuvier coined the term *Bimanes* (two hands). Each of them called apes *Quadrumanes* (four hands). Johann Friedrich Blumenbach, *Handbuch der Naturgeschichte* (Göttingen, 1779), 57–9; Georges Cuvier, *Le règne animal distribué d'après son organisation* (Paris, 1817), vol. i.

16. Georges-Louis Leclerc, Comte de Buffon, *Histoire naturelle, générale et particulière* (Paris, 1749–67) i. 38–40. The author of the article 'Mammifères' in the *Dictionnaire classique d'histoire naturelle* noted that in this period, it was commonly thought that male horses had no teats and consequently that mammae were not a universal characteristic of mammals (Paris, 1826), 10: 74. As John Lyon and Phillip Sloan have pointed out, Buffon may have been thinking of the stallion. Stallions have no teats and usually have inconspicuous rudimentary mammary glands, but even these are not always present. J. Lyon and P. Sloan (eds. and trans.) *From Natural History to the History of Nature: Readings from Buffon and His Critics* (Notre Dame, Ind., 1981), 94, n. 8.

17. Buffon, *Histoire naturelle*, i. 38–40. See also Phillip Sloan, 'The Buffon–Linnaeus Controversy', *Isis*, 67 (1976), 356–75; and James Larson, 'Linné's French Critics', Broberg, *Linnaeus*, 67–79.

18. Henri de Blainville, 'Prodrome: D'une nouvelle distribution systématique du règne animal', *Journal de physique*, 83 (1816), 246. See also Toby Appel, 'Henri de Blainville and the Animal Series: A Nineteenth-Century Chain of Being', *Journal of the History of Biology*, 13 (1980), 291–319, esp. 301.

19. John Hunter, *Essays and Observations on Natural History, Anatomy, Physiology, Psychology, and Geology*, Richard Owen (ed.), (London, 1861), i. 25.

20. Mamma meaning breast first appeared in English in 1579. Henry Skinner, *The Origin of Medical Terms* (Baltimore, Md., 1949), 223.
21. Blumenbach claimed that male hamsters and dormice do not have breasts but did not for this reason remove them from the class of mammals. *Handbuch der Naturgeschichte*, 46.
22. *Encyclopédie, ou Dictionnaire raisonné des sciences, des arts et des métiers* (Paris, 1751–65), vol. 10, *s.v.* 'Mamelle'.
23. Aristotle, *Historia Animalium*, in *The Works of Aristotle*, trans. D'arcy Thompson (Oxford, 1910), iv. 522a.
24. Buffon, *Histoire naturelle*, ii. 543.
25. Hunter, *Essays and Observations on Natural History*, 238–9.
26. Erasmus Darwin, *Zoonomia; or, The Laws of Organic Life* (London, 1796), i. 512.
27. G. Gegenbauer cited in Charles Darwin, *The Descent of Man and Selection in Relation to Sex* (1871; rpt. edn., London, 1913), 251, n. 29.
28. Darwin, *Descent of Man*, 249–53. Darwin cited Royer's *Origine de l'homme et des sociétés* (Paris, 1870). On Royer, see Joy Harvey, '"Strangers to Each Other": Male and Female Relations in the Life and Work of Clémence Royer', Pnina G. Abir-Am and Dorinda Outram (eds.), *Uneasy Careers and Intimate Lives: Women in Science, 1789–1979* (New Brunswick, NJ, 1987), 147–71.
29. Stephen Jay Gould, 'Freudian Slip', *Natural History* (Feb. 1987), 14–19.
30. Ronald King in Robert Thornton, *The Temple of Flora* (1799; rpt. edn., Boston, 1981), 9. Linnaeus sometimes named new genera after friends and colleagues, intending to suggest a spiritual likeness between the individual and the plant or animal in question; Benjamin Jackson, *Linnaeus* (London, 1923), 278. Linnaeus also ranked his colleagues as 'Officers in Flora's Army' according to his evaluation of their scientific merit. His list was headed by 'General Linnaeus'; the lowliest rank was assigned to Siegesbeck. Goerke, *Linnaeus*, 108.
31. Cuvier, *Le règne animal*, i. 76.
32. The other characteristics are: a jaw articulation formed by the squamosal and the dentary; a chain of three bones, malleus, incus, and stapes connecting the tympanic membrance to the inner ear, the presence of hair or fur; the left aortic arch in the systemic arch; and cheek teeth with divided roots. D. M. Kermack and K. A. Kermack, *The Evolution of Mammalian Characters* (London, 1984), vii; see also T. S. Kemp, *Mammal-like Reptiles and the Origin of Mammals* (London, 1982).
33. Scott Atran, *Cognitive Foundations of Natural History: Towards an Anthropology of Science* (Cambridge, 1990), 316, nn. 23–4.
34. Stephen Jay Gould, 'A Quahog Is a Quahog', in S. J. Gould (ed.), *The Panda's Thumb: More Reflections in Natural History* (New York, 1980), 204–7.
35. The cultural significance of the breast and mother's milk is a vast and as yet insufficiently studied topic; here I want to touch on only those aspects relevant to Linnaeus's work. Marina Warner's *Alone of All Her Sex: The Myth and the Cult of the Virgin Mary* (New York, 1976) and her *Monuments and Maidens: The Allegory of the Female Form* (London, 1985) along with Caroline Bynum's *Jesus as Mother: Studies in the Spirituality of the High Middle Ages* (Berkeley, Calif., 1982) are extremely helpful, although they focus primarily on the Middle Ages.
36. See Lynn Hunt, *Politics, Culture, and Class in the French Revolution* (Berkeley, Calif., 1984), esp. pt. 1; also Warner, *Monuments and Maidens*, chs. 12–13.

37. Linnaeus saw reason as the principle characteristic distinguishing humans from other animals. In the preface to his *Fauna Svecica* (1746), he called reason 'the most noble thing of all' that places humans above all others. See also H. W. Janson, *Apes and Ape Lore in the Middle Ages and the Renaissance* (London, 1952), 74–5.

38. Plato, *Timaeus*, 91c. Plato seemed uncertain whether woman should be classed with brute beasts or rational beings. Ian Maclean, *The Renaissance Notion of Woman: A Study in the Fortunes of Scholasticism and Medical Science in European Intellectual Life* (Cambridge, 1980), 31.

39. Warner, *Alone of All Her Sex*, 194.

40. Carl Linnaeus, 'Nutrix noverca', respondent F. Lindberg (1752), *Amoenitates academicae* (Erlangen, 1787), iii. 262–3. Goats and other animals were used to suckle syphilitic children in foundling hospitals in the eighteenth century or when there was a shortage of human nurses. Valerie Fildes, *Wet Nursing: A History from Antiquity to the Present* (Oxford, 1988), 147.

41. Mervyn Levy, *The Moons of Paradise: Some Reflections on the Appearance of the Female Breast in Art* (London, 1962), 55.

42. Hermann Ploss, Max Bartels, and Paul Bartels, in Eric John Dingwall (ed.), *Woman: An Historical, Gynaecological and Anthropological Compendium* (German edn., 1885; St. Louis, Mo., 1936), iii. 211.

43. William Godwin, *Memoirs of the Author of a Vindication of the Rights of Woman* (London, 1798), 183.

44. Petrus Camper did not explain why he used a female figure to illustrate the art of transforming 'a quadruped into the human figure'. *The Works of the Late Professor Camper: On the Connexion between the Science of Anatomy and the Arts of Drawing, Painting, Statuary, & c.*, T. Cogan, trans. (London, 1794), plate 7, fig. 13.

45. Carolyn Merchant, *The Death of Nature: Women, Ecology, and the Scientific Revolution* (San Francisco, 1980).

46. Charles Cochin and Hubert François Gravelot, *Iconologie par figures; ou, Traité complet des allégories, emblèmes & c* (1791; rpt. edn., Geneva, 1972), *s.v.* 'Nature'. Erasmus Darwin also portrayed 'Nature' as multi-breasted in *The Temple of Nature* (London, 1803), frontispiece.

47. Erich Neumann, *Grosse Mutter: Der Archetyp des grossen Weiblichen* (Zürich, 1956), 128.

48. Warner, *Alone of All Her Sex*, 192, 200; *Monuments and Maidens*, 283.

49. Bynum, *Jesus as Mother*, 115. See also Erwin Panofsky (ed.), *Abbot Suger on the Abbey Church of St.-Denis and Its Art Treasures* (Princeton, NJ, 1946), 30–1.

50. Warner, *Alone of All Her Sex*, 194.

51. Johann Wolfgang Goethe, *Faust: Eine Tragödie* (1808–32; rpt. edn., Munich, 1962), 19.

52. Warner, *Monuments and Maidens*, 281.

53. John Pinkerton, *A General Collection of the Best and Most Interesting Voyages and Travels in All Parts of the World* (London, 1808), ii. 194.

54. Mary Lindemann, 'Love for Hire: The Regulation of the Wet-Nursing Business in Eighteenth-Century Hamburg', *Journal of Family History*, 6 (1981), 382.

55. Linnaeus, 'Nutrix noverca'. Translated by Gilbert as 'La nourrice marâtre, ou dissertation sur les suites funestes du nourrissage mercénaire', *Les chef-d'oeuvres de Monsieur de Sauvages*, (Lyons, 1770), ii. 215–44.

56. Linnaeus, 'Nutrix noverca', 258.

57. Jean-Jacques Rousseau, *Emile: ou, De l'education* (1762), in Bernard Gegnebin and Marcel Raymond (eds.), *Œuvres complètes* (Paris, 1959–69), iv. 254–64.
58. Mary Jacobus, 'Incorruptible Milk: Breast-feeding and the French Revolution', Sara Melzer and Leslie Rabine (eds.), *Rebel Daughters: Women and the French Revolution* (New York, 1992), 62.
59. Lindemann, 'Love for Hire', 391.
60. See Hunt, *Politics, Culture, and Class in the French Revolution*, ch. 2, 3.
61. Cited in Darline Gay Levy, Harriet Bransom Applewhite, and Mary Durham Johnson (eds.), *Women in Revolutionary Paris, 1789–1795* (Urbana, Ill., 1979), 219.
62. Lynn Hunt, *The Family Romance of the French Revolution* (Berkeley, Calif., 1992), 151–91, esp. 153–5.
63. Londa Schiebinger, 'The Private Life of Plants: Sexual Politics in Carl Linnaeus and Erasmus Darwin', in Marina Benjamin (ed.), *Science and Sensibility: Gender and Scientific Enquiry 1780–1945* (Oxford, 1991), 121–43.

10

Language and Ideology in Evolutionary
Theory: Reading Cultural Norms into
Natural Law

Evelyn Fox Keller

In the mid-twentieth century, biology became a 'mature science', that is, it succeeded, finally, in breaking through the formidable barrier of 'life' that had heretofore precluded it from fully joining the mechanico-reductive tradition of the physical sciences. For the first time in history, the 'secret of life' could credibly be claimed to have been solved. Living beings—presumably including man along with the rest of the animal kingdom—came to be understood as (mere) chemical machines.

But the chemical machine that constitutes a living organism is unlike machines of the eighteenth and nineteenth centuries. It is not a machine capable only of executing the purposes of its maker, that is, man, but a machine endowed with its own purpose. In short, it is a machine of the twentieth century, a cybernetic machine *par excellence*: absolutely autonomous, capable of constructing itself, maintaining itself, and reproducing itself. As Jacques Monod has explained:

... the entire system is totally, intensely conservative, locked into itself, utterly impervious to any 'hints' from the outside world. Through its properties, by the microscopic clockwork function that establishes between DNA and protein, as between organism and medium, an entirely one-way relationship, this system obviously defies any 'dialectical' description. It is not Hegelian at all, but thoroughly Cartesian: the cell is indeed a machine. (1972: 110–11)

The purpose—the sole purpose—of this machine is its own survival and reproduction, or perhaps more accurately put, the survival and reproduction of the DNA programming and 'dictating' its operation. In Richard Dawkins's terms, an organism is a 'survival machine', a 'lumbering robot' constructing to house its genes, those 'engines of self-preservation' that have as their primary property that of being inherently 'selfish'. They are 'sealed off from the outside world, communicating with it by tortuous indirect routes, manipulating it by remote control. They are in you and in me; they created us, body and mind; and their preservation is the ultimate rationale for our existence' (1976: 21). With this description, man himself has become a machine, but perhaps it might alternatively be said that the machine itself has become man.

The general question is this: to what extent can our contemporary scientific description of animate forms, culminating in the description of man as a chemical machine, be said to be strictly technical, and to what extent does it actually encode particular conceptions of man—conceptions that derive not so much from a technical domain as from a social, political, and even psychological domain? Have animate, even human, forms finally been successfully deanimated and mechanized, or have their mechanical representations themselves been inadvertently animated, subtly recast in particular images of man?

I suggest that traces of such images might be found in virtually all scientific representations of nature, but they are perhaps especially conspicuous in our descriptions of the evolution of animate forms— even in those representations that make the greatest claims to conceptual neutrality. It is no secret that evolutionary biology has provided a particularly fertile field for those who seek to demonstrate the impact of social expectations on scientific theory. Indeed, it might be said that it is precisely for this reason that modern evolutionary theorists have sought so strenuously to place their discipline on firm scientific footing. Population genetics and mathematical ecology are the two subdisciplines that have been constructed to meet this need— to provide a rigorous substructure for all of evolutionary biology. The general methodological assumption that underlies both of these subdisciplines can be described as atomic individualism, that is, the assumption that a composite property of a system both can and should be represented by the aggregation of properties inhering in the individual atoms constituting that system, appropriately modified by their pairwise or higher order interactions.[1]

As is conventional in biological discourse, I take the individual atom to be, alternatively, the organism or the gene. I shall, therefore, focus on the particular attributes customarily assumed to characterize the basic unit of analysis, the individual organism or gene.

But my focus will be on the practice rather than the principle of atomic individualism in evolutionary theory. Others have argued for an ideological load in the very assumptions of this methodological orientation (see e.g. Wimsatt, Dupré, but here I wish to bracket such questions and focus instead on the lack of neutrality in its actual applications. In particular, I claim that properties of the 'individual' that are generally assumed to be necessary are in fact contingent, drawn not from nature but from our own social and psychosocial heritage. More specifically, I will argue that much of contemporary evolutionary theory relies on a representation of the 'individual'—be it the organism or the gene—that is cast in the particular image of man we might call the 'Hobbesian man': simultaneously autonomous and oppositional, connected to the world in which it finds itself not by the promise of life and growth but primarily by the threat of death and loss, its first and foremost need being the defence of its boundaries. In psychological terms, we might say that such an individual betrays an idealized conception of autonomy: one that presupposes a radical conception of self and that simultaneously attributes to the relation between self and other an automatic negative valence, a relation, finally, not so much of independence as of dynamic opposition.

I claim that this psychosocial load is carried into evolutionary theory not by explicit intention but by language—by tacit linguistic conventions that privilege the autonomy of the individual at the expense of biologically constitutive interactions and, at the same time, obscure the logical distinction between autonomy and opposition. In this, they support the characterization of the biological individual as somehow 'intrinsically' competitive, as if autonomy and competition were semantically equivalent, collapsed into one by that fundamentally ambiguous concept, self-interest. Accordingly, it is the language of autonomy and opposition in contemporary evolutionary theory that is the specific object of my concern.

DISCOURSE OF SELF AND OTHER

I begin with a relatively accessible example of actual confusion between autonomy and opposition that is found not in the theoretical literature *per se* but in a more general genre of scientific discourse, at once popularizing and prescriptive. Here the focus is not so much on the independence of one individual from another, of self from other, as on the independence of the most abstract other from self—of nature from man. Accordingly, the negative value that tacitly accrues to this relation attaches not so much to the self as to the other, that is, to nature.

With Darwin, evolutionary biology joined a tradition already well established in the physical sciences—a tradition that teaches that the laws of nature are, in Steven Weinberg's words, 'as impersonal and free of human values as the rules of arithmetic' (Weinberg 1974). But this rhetoric goes beyond impersonality: nature becomes uncaring and 'hostile', traits that are impersonal in a quite personal sense. To illustrate this tendency, consider, for example, Weinberg's own elaboration of his message:

It is almost irresistible for humans to believe that we have some special relation to the universe, that human life is not just a more-or-less farcical outcome of a chain of accidents reaching back to the first three minutes. . . . It is very hard to realize that this all is just a tiny part of an overwhelmingly hostile universe. (Midgley 1985: 88)

In much the same vein, Jacques Monod writes,

If he accepts this message in its full significance, man must at last wake out of his millenary dream and discover his total solitude, his fundamental isolation, he must realize that, like a gypsy, he lives on the boundary of an alien world, a world that is deaf to his music, and as indifferent to his hopes as it is to his suffering or his crimes. (Monod 1972: 2)

The world we must steel ourselves to accept is a world of 'uncaring emptiness', a 'frozen universe of solitude' (Monod 1972: 173). The natural world from which animism has been so carefully expunged has become not quite neutral but 'empty', 'frozen', 'overwhelmingly hostile', and 'terrifying'.

For the record, though, it was a poet, not a scientist, who gave us our most familiar metaphor conflating an absence of benevolence in nature with 'overwhelming hostility'. A nature that does not care for us becomes indeed a nature 'red in tooth and claw'—callous, brutal, even murderous. It is a nature that cries. 'I care for nothing; all shall go'.

Mary Midgley suggests that such residual animism properly belongs to what she calls 'the drama of parental callousness':

First, there is the tone of personal aggrievement and disillusion, which seems to depend . . . on failure to get rid of the animism or personification which (these scientists) officially denounce. An inanimate universe cannot be hostile. . . . Only in a real, conscious human parent could uncaringness equal hostility. . . . Certainly if we expect the non-human world around us to respond to us as a friendly human would, we shall be disappointed. But this does not put it in the position of a callously indifferent human. (1985: 87)

Midgley's explanation is persuasive—perhaps precisely because the slippage between an indifferent and a hostile nature so clearly does denote a logical error, once pointed out. But a similar problem surfaces in another set of contexts as well—where the move from a neutral to a negative valence in the conceptualization of self–other relations is less evidently a simple 'mistake'. Here it is not nature but the individual organism, the first rather than the second term of the self–other dichotomy, whom the insistently unsentimental biologist taxes with hostility.

I am referring in particular to the tradition among evolutionary biologists that not only privileges the individual descriptively but that also, in the attempt to locate all relevant causal dynamics in the properties intrinsic to the individual, tends to attribute to that individual not simply autonomy but an additional 'intrinsic' competitive bent, as if independence and competition were inseparable traits. The very same move that defines self-interest and altruism as logically opposed makes independence virtually indistinguishable from competition. Michael Ghiselin is one of the most extreme representatives of this position and provides some particularly blatant examples of the rhetorical (and conceptual) conflation I am speaking of. To dramatize his position, he concludes:

The economy of nature is competitive from beginning to end. . . . No hint of genuine charity ameliorates our vision of society, once sentimentalism has been laid aside. . . . Given a full chance to act for his own interest, nothing but expediency will restrain [an organism] from brutalizing, from maiming, from murdering—his brother, his mate, his parent, or his child. Scratch an 'altruist' and watch a 'hypocrite' bleed. (1974: 247)

Of course, Ghiselin's language is *intended* to shock us—but only to underscore his thesis. In this effort, he is counting on our acceptance, as readers, first, of the rule of self-interest as logically equivalent to the absence of altruism or charity, and second, of competitive exploitation as a necessary concomitant of self-interest. Our usual willingness to accept these assumptions, or rather, to allow them to pass unnoticed, is itself a measure of the inaccessibility of a domain where self-interest and charity (or altruism) conjoin and, correlatively, of a distinction between self-interest and competition. Unlike the previous example, where no one, if pressed, would say that nature 'really is' hostile, Ghiselin's assumptions do seem to accord with the way things 'really are'. Because the difference between self-interest and competition is less obvious to most of us than the difference between impersonality in nature and hostility, the problem here is much more difficult. In other words, before we can invoke a psychosocial explanation of the conceptual conflation between radical individualism and competition, we need first to *see* them as different.

We can do this best by turning our attention from a prescriptive discourse that aims to set ground rules for evolutionary theory to an examination of uses of the same language in the actual working out of particular theoretical strategies. I want to look, therefore at the technical uses of the language of individualism in mathematical ecology and population genetics—in the first case, on the language of competition, and, in the second, on the language of reproductive autonomy. In particular, I want to show how certain conventional interchanges, or trade-offs, between technical and colloquial language cast a blanket of invisibility, or rather, of unspeakability, over certain distinctions, categories, and questions. It is, I suggest, precisely through the maintenance of such an aura of unspeakability that social, psychological, and political expectations generally exert their influence, through language, on the actual structure and content of scientific theory.

COMPETITION AND CO-OPERATION IN MATHEMATICAL ECOLOGY

One problem I want to examine arises in the systematic neglect of co-operative (or mutualist) interactions and the corresponding privileging of competitive interactions evident throughout almost the entire history of mathematical ecology. When we ask practitioners in the field for an explanation of this historical disinterest in mutualist interactions, their response is usually one of puzzlement—not so much over the phenomenon as over the question, How else could it, realistically, be? Yes, of course, mutualist interactions occur in nature, but they are not only rare, they are necessarily secondary.

Often it is assumed that they are in the service of competition: such phenomena have at times actually been called 'co-operative competition'. Indeed, the expectation of most workers in the field that competition is both phenomenologically primary *and* logically prior is so deeply embedded that the very question has difficulty getting airspace: there is no place to put it. My question thus becomes, What are the factors responsible for the closing-off of that space?

Part of the difficulty in answering this question undoubtedly stems from the massive linguistic confusion in conventional use of the term *competition*. One central factor can be readily identified, however, and that is the recognition that, in the real world, resources *are* finite and hence ultimately scarce. To most minds, scarcity automatically implies competition, both in the sense of 'causing' competitive behaviour and in the sense of constituting, in itself, a kind of *de facto* competition, independent of actual interactions between organisms. Indeed, so automatic is the association between scarcity and competition that, in modern ecological usage, competition has come to be defined as the simultaneous reliance of two individuals, or two species, on an essential resource that is in limited supply (see, e.g. Mayr 1963: 43). Since the scarcity of resources can itself hardly be questioned, such a definition lends to competition the same a priori status.

This technical definition of competition was probably first employed by Volterra, Lotka, and Gause in their early attempts to provide a mathematical representation of the effects of scarcity on the population growth of 'interacting' species, but it soon came to be embraced by a wider community of evolutionary biologists and ecologists—partly, at least, to neutralize the discourse and so bypass the charge of ideologically laden expectations about (usually animal) behaviour, in fact freeing the discourse of any dependence on how organisms actually behave in the face of scarcity. The term *competition* now covered apparently pacific behaviour just as well as aggressive behaviour, an absurdity in ordinary usage but protected by the stipulation of a technical meaning. As Ernst Mayr explains,

To certain authors ever since [Darwin], competition has meant physical combat, and, conversely, the absence of physical combat has been taken as an indication of the absence of competition. Such a view is erroneous. . . . The relative rarity of overt manifestations of competition is proof not of the insignificance of competition, as asserted by some authors, but, on the contrary, of the high premium natural selection pays for the development of habits or preferences that reduce the severity of competition. (1963: 42–3)

Paul Colinvaux goes one step farther, suggesting that 'peaceful coexistence' provides a better description than any 'talk of struggles for survival'. 'Natural selection designs different kinds of animals and plants so that they *avoid* competition. A fit animal is not one that fights well, but one that avoids fighting altogether' (1978: 144).

But how neutral in practice is the ostensibly technical use of competition employed both by Mayr and Colinvaux? I suggest two ways in which, rather than bypassing ideological expectations, it actually preserves them, albeit in a less visible form—a form in which they enjoy effective immunity from criticism. So as not to be caught in the very trap I want to expose, let me henceforth denote competition in the technical sense as 'Competition' and in the colloquial sense (of actual contest) as 'competition'.

The first way is relatively straightforward. The use of a term with established colloquial meaning in a technical context permits the simultaneous transfer and denial of its colloquial connotations. Let me offer just one example: Colinvaux's own description of Gause's original experiments that were designed to study the effect of scarcity on interspecific dynamics—historically, the experimental underpinning of the 'competitive exclusion coexistence'. He writes,

No matter how many times Gause tested [the paramecia] against each other, the outcome was always the same, complete extermination of one species. . . . Gause could see this deadly struggle going on before his eyes day after day and always with the same outcome. . . . What we [might have] expected to be a permanent struggling balance in fact became a pogrom. (1978: 142)

Just to set the record straight, these are not 'killer' paramecia but perfectly ordinary paramecia—minding their own business, eating and dividing, or not, perhaps even starving. The terms *extermination, deadly struggle,* and *pogrom* refer merely to the simultaneous dependence of two species of a common resource. If, by chance, you should have misinterpreted and taken them literally, to refer to overt combat, you would be told that you had missed the point. The Lotka–Volterra equations make no such claims; strictly speaking, they are incompatible with an assumption of overt combat; the competitive exclusion principle merely implies an avoidance of conflict. And yet the description of such a situation, only competitive in the technical sense, slips smoothly from 'Competition' to genocide—much as we saw our neo-Tennysonians slip from impersonality to heartless rejection.

The point of this example is not to single out Colinvaux (which would surely be unfair) but to provide an illustration of what is, in fact, a rather widespread investment of an ostensibly neutral technical term with a quite different set of connotations associated with its colloquial meaning. The colloquial connotations lead plausibly to one set of inferences and close off others, while the technical meaning stands ready to disclaim responsibility if challenged.[2]

The second and more serious route by which the apparently a priori status of competition is secured can be explored through an inquiry into the implicit assumptions about resource consumption that are here presupposed and the aspects of resource consumptions that are excluded. The first presupposition is that a resource can be defined and quantitatively assessed independent of the organism itself; and the second is that each organism's utilization of this resource is independent of other organisms. In short, resource consumption is here represented as a zero-sum game. Such a representation might be said to correspond to the absolutely minimal constraint possible on the autonomy of each individual, but it is a constraint that has precisely the effect of establishing a necessary link between self-interest and competition. With these assumptions, apparently autonomous individuals are in fact bound by a zero-sum dynamic that guarantees not quite an absence of interaction but the inevitability of a purely competitive interaction. In a world in which one organism's dinner necessarily means another's starvation, the mere consumption of resources has a kind of *de facto* equivalence to murder: individual organisms are locked into a life and death struggle not by virtue of their direct interactions but merely by virtue of their existence in the same place and time.

It is worth noting that the very same (Lotka–Volterra) equations readily accommodate the replacement of competitive interactions by co-operative ones and even yield a stable solution. This fact was actually noted by Gause himself as early as 1935 (Gause and Witt 1935) and has been occasionally rediscovered since then, only to be, each time, reforgotten by the community of mathematical ecologists. The full reasons for such amnesia are unclear, but it suggests a strong prior commitment to the representation of resource consumption as a zero-sum dynamic—a representation that would be fatally undermined by the substitution (or even addition) of co-operative interactions.

Left out of this representation are not only co-operative interactions but *any* interactions between organisms that affect the individual's need and utilization of resources. Also omitted are all those interactions between organism and environment which interfere with the identification and measurement of a resource independently of the properties of the organism. Richard Lewontin (1982) has argued that organisms 'determine what is relevant' in their environment—what, for example, *is* a resource—and actually 'construct' their environment. But such interactions, either between organisms or between organism and environment, lead to pay-off matrices necessarily more complex than those prescribed by a zero-sum dynamic—pay-off matrices that, in turn, considerably complicate the presumed relation between self-interest and competition, if they do not altogether undermine the very meaning of self-interest.

Perhaps the simplest example is provided by the 'prisoner's dilemma'. But even here, where the original meaning of self-interest is most closely preserved, Robert Axelrod has shown that under conditions of indefinite reiterations, a ('tit-for-tat') strategy is generally better suited to self-interest than are more primitive competitive strategies.

Interactions that effectively generate new resources, or either increase the efficiency of resource utilization or reduce absolute requirement, are more directly damaging to the very principle of self-interest. These are exactly the kinds of interactions that are generally categorized as special cases: as 'mutualist', 'co-operative', or 'symbiotic' interactions. Finally, interactions that affect the birth rate in ways not mediated by scarcity of resources, for example, sexual reproduction, are also excluded by this representation. Perhaps the most important of these omissions for interspecific dynamics is that of mutualist interactions, while for intraspecific dynamics, I would point to sexual reproduction, a fact of life that potentially undermines the core assumptions of radical individualism. In the last few years, there has been a new wave of interest in mutualism among not only dissident but even a few mainstream biologists, and numerous authors are hard at work redressing the neglect of previous years.[3] But in the sixty years in which the Lotka–Volterra equations have reigned as the principal, if not the only, model of interspecific population dynamics—even in the more genial climate of recent years—the omission of sexual reproduction from this model has scarcely been noted.

This omission, once recognized, takes us beyond the question of selective biases in admissible or relevant interactions between organisms. It calls into question the first and most basic assumption for the methodology of individualism in evolutionary theory, namely, that intrinsic properties of individual organisms are primary to any description of evolutionary phenomena[4] To examine this argument, let us turn from mathematical ecology to population genetics, that branch of evolutionary theory that promises to avoid the practical difficulties of selective focus on certain interactions by excluding the entire question of competitive or co-operative interactions from its domain. In other words, traditional population genetics addresses neither interactions between organisms nor limitations in resources; it effectively assumes populations at low density with infinite resources.

However, one last problem with the language of competition must be noted lest it carry over into our discussion of individual autonomy in population genetics: the widespread tendency to extend the sense of 'competition' to include not only the two situations we distinguished earlier (i.e. conflict and reliance on a common resource) but also a third situation[5] where there is no interaction at all, where 'competition' denotes an operation of *comparison* between organisms (or species) that requires no juxtaposition in nature, only in the biologist's own mind. This extension, where 'competition' can cover all possible circumstances of relative viability and reproductivity, brings with it, then, the tendency to equate competition with natural selection itself.

Darwin's own rhetorical equation between natural selection and the Malthusian struggle for existence surely bears some responsibility for this tendency. But contemporary readers of Darwin like to point out that he did try to correct the misreading his rhetoric invited by explaining that he meant the term *struggle* in 'a large and metaphoric sense', including, for example, that of the plant on the edge of the desert: competition was only one of the many meanings of struggle for Darwin. Others have been even more explicit on this issue, repeatedly noting the importance of distinguishing natural selection from 'a Malthusian dynamic'. Lewontin, for one, has written.

Thus, although Darwin came to the idea of natural selection from consideration of Malthus' essay on overpopulation, the element of competition between organisms for a resource in short supply is not integral to the argument. Natural selection occurs even when two bacterial strains are growing logarithmically in an excess of nutrient broth if they have different division times. (1970: 1)

However, such attempts—by Lewontin, and earlier and more comprehensively, by L. C. Birch (1957)—to clarify the distinction between natural selection and competition (what Engels called 'Darwin's mistake') have done little to stem the underlying conviction that the two are somehow the same. In a recent attempt to define the logical essence of 'the Darwinian dynamic', Bernstein *et al.* (1983) freely translate Darwin's 'struggle for survival' to 'competition through resource limitation' (192), thereby claiming for competition the status of a 'basic component' of natural selection. Even more recently, George Williams describes a classic example of natural selection in the laboratory as a 'competition experiment', a 'contest' between a mutant and normal allele, in which he cites differential fecundity as an example of 'the competitive interactions among individual organisms' that cause the relative increase in one population (1986: 114–15).

At issue is not whether overtly competitive behaviour or more basic ecological scarcity is the rule in the natural world; rather, it is whether or not such a question can even be asked. To the extent that distinctions between competition and scarcity, on the one hand, and between scarcity and natural selection, on the other, are obliterated from our language and thought, the question itself becomes foreclosed. As long as the theory of natural selection is understood as a theory of competition, confirmation of one is taken to be confirmation of the other, despite their logical (and biological) difference.

While this clearly raises problems about the meaning of confirmation, my principal concern is with the dynamics by which such an oversight or confusion is sustained in the theory and practice of working biologists—with the internal conventions that render it effectively resistant to correction. Dynamics similar to those in the language of competition can also be seen in the language of reproductive autonomy, especially as employed in the theory and practice of population biology.

THE PROBLEM OF SEXUAL REPRODUCTION

In much of the discourse on reproduction, it is common to speak of the 'reproduction of an organism'—as if reproduction is something an individual organism does; as if an organism makes copies of itself, by itself. Strictly speaking, of course, such language is appropriate only to asexually reproducing populations since, as every biologist knows, sexually reproducing organisms neither produce copies of themselves nor produce other organisms by themselves. It is a striking fact, however, that the language of individual reproduction, including such correlative terms as *an individual's offspring* and *lineage*, is used throughout population biology[6] to apply indiscriminately to both sexually and asexually reproducing populations. While it would be absurd to suggest that users of such language are actually confused about the nature of reproduction in the organisms they study (e.g. calculations of numbers of offspring per organism are always appropriately adjusted to take the mode of reproduction into account), we might none the less ask, what functions, both positive and negative, does such manifestly peculiar language serve? And what consequences does it have for the shape of the theory in which it is embedded?

I want to suggest, first, that this language, far from being inconsequential, provides crucial conceptual support for the individualist programme in evolutionary theory. In particular, my claim is that the starting assumption of this programme—that is, that individual properties are primary—depends on the language of individual reproduction for its basic credibility.[7] In addition, I would argue that, just as we saw with the language of competition, the language of individual reproduction, maintained as it is by certain methodological conventions, both blocks the perception of problems in the evolutionary project as presently conducted and, simultaneously, impedes efforts to redress those difficulties that can be identified.

The problems posed for evolutionary theory by sexual reproduction and Mendelian genetics are hardly new, and indeed, the basic theory of population genetics originates in the formulation of a particular method (i.e. the Hardy–Weinberg calculus) designed to solve these problems. The Hardy–Weinberg calculus (often referred to as 'bean-bag' genetics) invoked an obviously highly idealized representation of the relation between genes, organisms, and reproduction, but it was one that accomplished a great deal. Most important, it provided a remarkably simple recipe for mediating between individuals and populations—a recipe that apparently succeeded in preserving the individualist focus of the evolutionists' programme. One might even say that it did so, perhaps somewhat paradoxically, by tacitly discounting individual organisms and their troublesome mode of reproduction. With the shift of attention from populations of organisms to well-mixed, effectively infinite, pools of genes, the gap between individual and population closed. Individual organisms, in this picture, could be thought of as mere bags of genes (anticipating Richard Dawkins's 'survival machines' (1976: 21))—the end-product of a reproductive process now reduced to genetic replication plus the random mating of gametes. Effectively bypassed with this representation were all the problems entailed by sexual difference, by the contingencies of mating and fertilization that resulted from the finitude of actual populations and, simultaneously, all the ambiguities of the term *reproduction* as applied to organisms that neither make copies of themselves nor reproduce by themselves. In short, the Hardy–Weinberg calculus provided a recipe for dealing with reproduction that left undisturbed—indeed, finally, reinforced—the temptation to think (and to speak) about reproduction as simply an individual process, to the extent, that is, that it was thought or spoken about at all.

In the subsequent incorporation of the effects of natural selection into the Hardy–Weinberg model, for most authors in population genetics, the contribution of reproduction to natural selection fell largely by the wayside. True, the basic calculus provided a ready way to incorporate at least part of the reproductive process, namely, the production of gametes; but in practice, the theoretical (and verbal) convention that came to prevail in traditional population genetics was to equate natural selection with differential survival and ignore fertility altogether. In other words, the Hardy–Weinberg calculus seems to have invited not one but two kinds of elision from natural selection—first, of all those complications incurred by sex and the contingency of mating (these, if considered at all, get shunted off under the label of sexual, rather than natural, selection),[8] and second, more obliquely, of reproduction *in toto*.

I want to suggest that these two different kinds of elision in fact provided important tacit support for each other. In the first case, the representation of reproduction as gametic production invited confidence in the assumption that, for calculating changes in gene frequency, differential reproduction, or fertility, was *like* differential survival and hence did not require separate treatment. And in the second case, the technical equation of natural selection with differential survival which prevailed for so many years, in turn, served to deflect attention away from the substantive difficulties invoked in representing reproduction as an individual process. The net effect has been to establish a circle of confidence, first, in the adequacy of the assumption that, despite the mechanics of Mendelianism, the individual remains both the subject and object of reproduction, and second, in the adequacy of the metonymic collapse of reproduction and survival in discussions of natural selection.

The more obvious cost of this circle surely comes from its second part. As a number of authors have recently begun to remind us, the equation between natural selection and differential survival fosters both the theoretical omission and the experimental neglect of a crucial component of natural selection. Perhaps even more serious is the cost in unresolved difficulties that this equation has helped obscure.

One such difficulty is the persistence of a chronic confusion between two definitions of individual fitness: one, the (average) net contribution of an individual of a particular genotype to the next generation, and the other, the geometric rate of increase of that particular genotype. The first refers to the contribution an individual makes to reproduction, while the second refers to the rate of production of individuals. In other words, the first definition refers to the role of the individual as subject of reproduction and the second to its role as object. The disparity between the two derives from the basic fact that, for sexually reproducing organisms, the rate at which individuals of a particular genotype are born is a fundamentally different quantity from the rate at which individuals of that genotype give birth—a distinction easily lost in a language that assigns the same term, *birth rate*, to both processes.

Beginning in 1962, a number of authors have attempted to call attention to this confusion (Moran 1962; Charlesworth 1970; Pollak and Kempthorne 1971; Denniston 1978), agreeing that one definition—the contribution a particular genotype makes to the next generation's population—is both conventional and correct, while the other (the rate at which individuals of a particular genotype are born) is not. Despite their efforts, however, the confusion persists.[9] In part, this is because their remains a real question as to what 'correct' means in this context or more precisely, as to which definition is better suited to the needs that the concept of fitness is intended to serve—in particular, the need to explain changes in the genotypic composition of populations. Given that need, we want to know not only which genotypes produce more but also the relative rate of increase of a particular genotype over the course of generations.

Not surprisingly, conflation of the two definitions of fitness is particularly likely to occur in attempts to establish a formal connection between the models of population genetics and those of mathematical ecology. Because all the standard models for population growth assume asexual reproduction, the two formalisms actually refer to two completely different kinds of populations: one of gametic pools and the other of asexually reproducing organisms. In attempting to reconcile these two theories, such a conflation is in fact required to finesse the logical gap between them. A more adequate reconciliation of the two formalisms requires the introduction of both the dynamics of sexual reproduction into mathematical ecology and a compatible representation of those dynamics into population genetics.

Counterintuitively, it is probably the second—the inclusion (in population genetics models) of fertility as a property of the mating type—that calls for the more substantive conceptual shifts. Over the last twenty years, we have witnessed the emergence of a considerable literature devoted to the analysis of fertility selection—leading at least some authors to the conclusion that 'the classical concept of individual fitness is insufficient to account for the action of natural selection' (Christiansen 1983: 75).

The basic point is that when fertility selection *is* included in natural selection, the fitness of a genotype, like the fitness of a gene (as argued by Sober and Lewontin 1982), is seen to depend on the context in which it finds itself. Now, however, the context is one determined by the genotype of the mating partner rather than by the complementary allele. A casual reading of the literature on fertility selection might suggest that the mating pair would be a more appropriate unit of selection than the individual, but the fact is that mating pairs do not reproduce themselves any more than do individual genotypes. As E. Pollak has pointed out, 'even if a superior mating produces offspring with a potential for entering a superior mating, the realization of this potential is dependent upon the structure of the population' (1978: 389). In other words, in computing the contribution of either a genotype or a mating pair to the next generation's population (of genotypes or mating pairs), it is necessary to take account of the contingency of mating: such a factor, measuring the probability that any particular organism will actually mate, incurs a frequency dependence that reflects the dependence of mating on the genotypic composition of the entire population.

Very briefly, the inclusion of a full account of reproduction in evolutionary theory necessitates the conclusion that natural selection operates simultaneously on many levels (gene, organism, mating pair, and group), not just under special circumstances, as others have argued, but as a rule. For sexually reproducing organisms, fitness in general is not an individual property but a composite of the entire interbreeding population, including, but certainly not determined by, genic, genotypic, and mating pair contributions. By undermining the reproductive autonomy of the individual organism, the advent of sex undermines the possibility of locating the causal efficacy of evolutionary change in individual properties. At least part of the 'causal engine' of natural selection must be seen as distributed throughout the entire population of interbreeding organisms.

My point is not merely to argue against the adequacy of the individualist programme in evolutionary theory but—like the point of my earlier remarks about competition—to illustrate a quite general process by which the particular conventions of language employed by a scientific community permit a tacit incorporation of ideology into scientific theory and, simultaneously, protect participants from recognition of such ideological influences. The net effect is to insulate the theoretical structure from substantive critical revision. In discussions of sexual reproduction, the linguistic conventions of individual reproduction—conventions embodying an ideological commitment to the a priori autonomy of the individual—both perpetuate that belief and promote its incorporation into the theory of evolutionary dynamics. At the same time, the conventional equation between natural selection and differential survival has served to protect evolutionary theory from problems introduced by sexual reproduction, thereby lending at least tacit support to the assumption of individual autonomy that gave rise to the language of individual reproduction in the first place. The result—now both of the language of autonomy and the language of competition—is to effectively exclude from the domain of theory those biological phenomena that do not fit (or even worse, threaten to undermine) the ideological commitments that are unspoken yet *in* language, built into science by the language we use in both constructing and applying our theories. In this way, through our inescapable reliance on language, even the most ardent efforts to rid natural law of cultural norms become subverted, and the machinery of life takes on not so much a life of its own as a life of our own. But then again, what other life could it have?

Notes

1. The basic claim of atomic individualism can be schematically expressed as follows:

$$X = \sum_i X_i + \sum_{ij} X_{ij} + \Sigma_{ijk} + \Sigma_{ijkl} X_{ijkl} + \dots .$$

(successive orders of interaction are represented by the terms xij, $xijk$, $xijkl$, etc.) The actual implementation of this methodology depends, however, on three implicit assumptions:

   i. The first term in the series is primary;
   ii. All relevant interactions are included in the subsequent summations; and finally, that

iii. The series converges (i.e. there are no unexpected effects from neglected higher-order terms).

Ultimately, it seems to me that the application of all three of these assumptions of evolutionary theory is subject to serious question. My particular focus here, however, is on the adequacy of the first two assumptions.

2. See Keller (1988) for a discussion of Hardin's use of the same slippage in arguing for the universality of the 'competitive exclusion principle' (1960).

3. Douglas Boucher (1985) has even suggested a new metaphor: in place of 'nature red in tooth and claw', he offers 'nature green in root and bloom'.

4. That is, it raises a question about the adequacy of the third assumption of my schematic account of the methodology of individualism—that in which the essential (or existential) autonomy of the individual organism is assumed.

5. Which is, in fact, the situation of population genetics.

6. Including both population genetics and mathematical ecology.

7. For example, in the absence of other organisms, the fitness of a sexually reproducing organism is, strictly speaking, zero.

8. Darwin originally introduced the idea of sexual selection—always in clear contradistinction to natural selection—in an effort to take account of at least certain aspects of reproductive selection. . . . In my view, the recent interest in sexual selection among sociobiologists is a direct consequence of the final, and complete, abandonment of the individual organism as a theoretical entity. Genetic selection theories, it could be said, complete the shift of attention away from organisms begun by the Hardy-Weinberg calculus. Sexual reproduction is a problem in this discourse only to the extent that individual organisms remain, somewhere, an important (even if shifting) focus on conceptual interest.

9. See Keller (1987) for details.

References

Axelrod, Robert (1984), *The Evolution of Cooperation* (New York: Basic Books).

Bernstein, H., H. C. Byerly, F. A. Hopf, R. A. Michod, and G. K. Vemulapalli (1983), 'The Darwinian Dynamic', *The Quarterly Review of Biology*, 58, 185–207.

Birch, L. C. (1957), 'Meanings of Competition', *American Naturalist*, 91, 5–18.

Boucher, Douglas, (1985), *The Biology of Mutualism*, (Oxford: Oxford University Press).

Charlesworth, B. (1970), 'Selection in Populations with Overlapping Generations, I, The Use of Malthusian Parameters in Population Genetics', *Theoretical Population Biology*, 1/3, 352–70.

Christiansen, F. B. (1984), 'The Definition and Measurement of Fitness', in B. Shorrocks, (ed.), *Evolutionary Ecology: B. E. S. Symposium 23* (Oxford and Boston: Blackwell Scientific Publications) 65–79.

Colinvaux, Paul (1978), *Why Big Fierce Animals are Rare* (Princeton: Princeton University Press).

Dawkins, Richard (1976), *The Selfish Gene* (Oxford: Oxford University Press).

Denniston, C. (1978), 'An Incorrect Definition of Fitness Revisited', *Annals of Human Genetics, Lond.* 42, 77–85.

Gause, G. F., and A. A. Witt (1935), 'Behavior of Mixed Populations and the Problem of Natural Selection', *American Naturalist,* 69, 596–609.

Ghiselin, Michael (1974), *The Economy of Nature and the Evolution of Sex* (Berkeley, Los Angeles, London: University of California Press).

Keller, Evelyn Fox (1987), 'Reproduction and the Central Project of Evolutionary Theory', *Biology and Philosophy* 2, 73–86.

— (1988), 'Demarcating Public from Private Values in Evolutionary Discourse', *Journal of the History of Biology,* 21/2, 195–211.

Lewontin, Richard (1970), The Units of Selection', *Annual Review of Ecology and Systematics,* 1–1B.

— (1982), 'Organism and Environment', in E. H. C. Plotkin (ed.), *Learning, Development, and Culture* (New York: John Wiley and Sons).

Mayr, Ernst (1963), *Animal Species and Evolution* (Cambridge: Harvard University Press).

Midgley, Mary (1985), *Evolution as a Religion* (London and New York: Methuen).

Monod, Jacques (1972), *Chance and Necessity* (New York: Random House).

Moran, P. A. P. (1962), 'On the Nonexistence of Adaptive Topographies', *Annals of Human Genetics,* 27, 383–93.

Pollak, E. (1978), 'With Selection for Fecundity the Mean Fitness Does Not Necessarily Increase', *Genetics,* 90, 383–9.

— and O. Kempthorne (1971), 'Malthusian Parameters in Genetic Populations, II, Random Mating Populations in Infinite Habitats', *Theoretical Population Biology,* 2, 357–90.

Sober, E., and R. Lewontin (1982), 'Artifact, Cause, and Genetic Selection', *Philosophy of Science,* 47, 157–80.

Tennyson, Alfred, *In Memoriam,* LV–LVI.

Weinberg, Steven (1974), 'Reflections of a Working Scientist', *Daedalus,* 103 (Summer) 33–46.

Williams, George (1986), 'Comments', *Biology and Philosophy,* 1/1, 114–22.

Nuclear Language and How We Learned to Pat the Bomb

Carol Cohn

My close encounter with nuclear strategic analysis started in the summer of 1984. I was one of forty-eight college teachers attending a summer workshop on nuclear weapons, strategic doctrine, and arms control that was held at a university containing one of the nation's foremost centres of nuclear strategic studies, and that was co-sponsored by another institution. It was taught by some of the most distinguished experts in the field, who have spent decades moving back and forth between academia and governmental positions in Washington. When at the end of the programme I was afforded the chance to be a visiting scholar at one of the universities' defence studies centre, I jumped at the opportunity.
. . .

I attended lectures, listened to arguments, conversed with defence analysts, interviewed graduate students throughout their training, obsessed by the question, 'How *can* they think this way?' But as I learned the language, as I became more and more engaged with their information and their arguments, I found that my own thinking was changing, and I had to confront a new question: How can *I* think this way? Thus, my own experience becomes part of the data that I analyse in attempting to understand not only how 'they' can think that way, but how any of us can.

This article is the beginning of an analysis of the nature of nuclear strategic thinking, with emphasis on the role of a specialized language that I call 'technostrategic'. I have come to believe that this language both reflects and shapes the American nuclear strategic project, and that all who are concerned about nuclear weaponry and nuclear war must give careful attention to language—with whom it allows us to communicate and what it allows us to think as well as say.
. . .

Reprinted with permission from *Bulletin of the Atomic Scientists* (June 1987).

Feminists have often suggested that an important aspect of the arms race is phallic worship; that 'missile envy', to borrow Helen Caldicott's phrase, is a significant motivating force in the nuclear buildup. I have always found this an uncomfortably reductionist explanation and hoped that observing at the centre would yield a more complex analysis. Still, I was curious about the extent to which I might find a sexual subtext in the defence professionals' discourse. I was not prepared for what I found.

I think I had naïvely imagined that I would need to sneak around and eavesdrop on what men said in unguarded moments, using all my cunning to unearth sexual imagery. I had believed that these men would have cleaned up their acts, or that at least at some point in a long talk about 'penetration aids', someone would suddenly look up, slightly embarrassed to be caught in such blatant confirmation of feminist analyses.

I was wrong. There was no evidence that such critiques had ever reached the ears, much less the minds, of these men. American military dependence on nuclear weapons was explained as 'irresistible, because you get more bang for the buck'. Another lecturer solemnly and scientifically announced, 'To disarm is to get rid of all your stuff.' A professor's explanation of why the MX missile is to be placed in the silos of the newest Minuteman missiles, instead of replacing the older, less accurate missiles, was 'because they're in the nicest hole—you're not going to take the nicest missile you have and put it in a crummy hole'. Other lectures were filled with discussion of vertical erector launchers, thrust-to-weight ratios, soft lay downs, deep penetration, and the comparative advantages of protracted versus spasm attacks—or what one military adviser to the National Security Council has called 'releasing 70 to 80 per cent of our megatonnage in one orgasmic whump'.[1]

But if the imagery is transparent, its significance may be less so. I do *not* want to assert that it somehow reveals what defence intellectuals are really talking about, or their motivations; individual motives cannot necessarily be read directly from imagery, which originates in a broader cultural context. The history of the atomic bomb project itself is rife with overt images of competitive male sexuality, as is the discourse of the early nuclear physicists, strategists, and members of the Strategic Air Command.[2] Both the military itself and the arms manufacturers are constantly exploiting the phallic imagery and promise of sexual domination that their weapons so conveniently suggest. Consider the following, from the June 1985 issue of *Air Force Magazine*: Emblazoned in bold letters across the top of a two-page advertisement for the AV-8B Harrier II—'Speak Softly and Carry a Big Stick'. The copy below boasts 'an exceptional thrust-to-weight ratio', and 'vectored thrust capability that makes the . . . unique rapid response possible'.

Another vivid source of phallic imagery is to be found in descriptions of nuclear blasts themselves. Here, for example, is one by journalist William Laurence, who was brought by the Army Air Corps to witness the Nagasaki bombing.

Then, just when it appeared as though the thing had settled down into a state of permanence, there came shooting out of the top a giant mushroom that increased the size of the pillar to a total of 45 000 feet. The mushroom top was even more alive than the pillar, seething and boiling in a white fury of creamy foam, sizzling upward and then descending earthward, a thousand geysers rolled into one. It kept struggling in an elemental fury, like a creature in the act of breaking the bonds that held it down.[3]

Given the degree to which it suffuses their world, the fact that defence intellectuals use a lot of sexual imagery is not especially surprising. Nor does it, by itself, constitute grounds for imputing motivation. The interesting issue is not so much the imagery's possible psychodynamic origins as how it functions—its role in making the work world of defence intellectuals feel tenable. Several stories illustrate the complexity.

At one point a group of us took a field trip to the New London Navy base where nuclear submarines are home-ported, and to the General Dynamics Electric Boat yards where a new Trident submarine was being constructed. The high point of the trip was a tour of a nuclear-powered submarine. A few at a time, we descended into the long, dark, sleek tube in which men and a nuclear reactor are encased underwater for months at a time. We squeezed through hatches, along neon-lit passages so narrow that we had to turn and press our backs to the walls for anyone to get by. We passed the cramped racks where men sleep, and the red and white signs warning of radioactive materials. When we finally reached the part of the sub where the missiles are housed, the officer accompanying us turned with a grin and asked if we wanted to stick our hands through a hole to 'pat the missile'. *Pat the missile?*

The image reappeared the next week, when a lecturer scornfully declared that the only real reason for deploying cruise and Pershing II missiles in Western Europe was 'so that our allies can pat them'. Some months later, another group of us went to be briefed at NORAD (the North American Aerospace Defense Command). On the way back, the Air National Guard plane we were on went to refuel at Offut Air Force Base, the Strategic Air Command head-quarters near Omaha, Nebraska. When word leaked out that our landing would be delayed because the new B-1 bomber was in the area, the plane became charged with a tangible excitement that built as we flew in our holding pattern, people craning their necks to try to catch a glimpse of the B-1 in the skies, and climaxed as we touched down on the runway and hurtled past it. Later, when I returned to the centre I encountered a man who, unable to go on the trip, said to me enviously, 'I hear you got to pat a B-1'.

What is all this patting? Patting is an assertion of intimacy, sexual possession, affectionate domination. The thrill and pleasure of 'patting the missile' is the proximity of all that phallic power, the possibility of vicariously appropriating it as one's own. But patting is not only an act of sexual intimacy. It is also what one does to babies, small children, the pet dog. The creatures one pats are small, cute, harmless—not terrifyingly destructive. Pat it, and its lethality disappears.

Much of the sexual imagery I heard was rife with the sort of ambiguity suggested by 'patting the missiles'. The imagery can be construed as a deadly serious display of the connections between masculine sexuality and the arms race. But at the same time, it can also be heard as a way of minimizing the seriousness of militarist endeavours, of denying their deadly consequences. A former Pentagon target analyst, in telling me why he thought plans for 'limited nuclear war' were ridiculous, said, 'Look, you gotta under-stand that it's a pissing contest—you gotta expect them to use everything they've got.' This image says, most obviously, that this is about competition for manhood, and thus there is tremendous danger. But at the same time it says that the whole thing is not very serious—it is just what little boys or drunk men do.

Sanitized abstraction and sexual imagery, even if disturbing, seemed to fit easily into the masculine world of nuclear war planning. What did not fit was another set of words that evoked images that can only be called domestic.

Nuclear missiles are based in 'silos'. On a Trident submarine, which carries 24 multiple-warhead nuclear missiles, crew members call the part of the sub where the missiles are lined up in their silos ready for launching 'the Christmas tree farm'. In the friendly, romantic world of nuclear weaponry, enemies 'exchange' warheads; weapons systems can 'marry up'. 'Coupling' is sometimes used to refer to the wiring between mechanisms of warning and response, or to the psychopolitical links between strategic and theatre weapons. The patterns in which a MIRVed missile's nuclear warheads land is known as a 'footprint'. These nuclear explosives are not dropped; a 'bus' 'delivers' them. These devices are called 'reentry vehicles', or 'RVs' for short, a term not only totally removed from the reality of a bomb but also resonant with the image of the recreational vehicles of the ideal family vacation.

. . .

Another set of images suggests men's desire to appropriate from women the power of giving life. At Los Alamos, the atomic bomb was referred to as 'Oppenheimer's baby'; at Lawrence Livermore, the hydrogen bomb was 'Teller's baby', although those who wanted to disparage Teller's contribution claimed he was not the bomb's father but its mother. In this context, the extraordinary names given to the bombs that reduced Hiroshima and Nagasaki to ash and rubble—'Little Boy' and 'Fat Man'—may perhaps become intelligible. These ultimate destroyers were the male progeny of the atomic scientists.

. . .

Although I was startled by the combination of dry abstraction and odd imagery that characterizes the language of defence intellectuals, my attention was quickly focused on decoding and learning to speak it. The first task was training the tongue in the articulation of acronyms.

Several years of reading the literature of nuclear weaponry and strategy had not prepared me for the degree to which acronyms littered all conversations, nor for the way in which they are used. Formerly, I had thought of them mainly as utilitarian. They allow you to write or speak faster. They act as a form of abstraction, removing you from the reality behind the words. They restrict communication to the initiated, leaving the rest both uncomprehending and voiceless in the debate.

But being at the centre revealed some additional, unexpected dimensions. First, in speaking and hearing, a lot of these terms are very sexy. A small supersonic rocket 'designed to penetrate any Soviet air defence' is called a SRAM (for short-range attack missile). Submarine-launched cruise missiles are referred to as 'slick'ems' and ground-launched cruise missiles are 'glick'ems'. Air launched cruise missiles are magical 'alchems'.

Other acronyms serve in different ways. The plane in which the president will supposedly be flying around above a nuclear holocaust, receiving intelligence and issuing commands for where to bomb next, is referred to as 'Kneecap' (for NEACP—National Emergency Airborne Command Post). Few believe that the president would really have the time to get into it, or that the communications systems would be working if he were in it—hence the edge of derision. But the very ability to make fun of a concept makes it possible to work with it rather than reject it outright.

In other words, what I learned at the programme is that talking about nuclear weapons is fun. The words are quick, clean, light; they trip off the tongue. You can reel off dozens of them in seconds, forgetting about how one might interfere with the next, not to mention with the lives beneath them. Nearly everyone I observed—lecturers, students, hawks, doves, men, and women—took pleasure in using the words; some of us spoke with a self-consciously ironic edge, but the pleasure was there none the less. Part of the appeal was the thrill of being able to manipulate an arcane language, the power of entering the secret kingdom. But perhaps more important, learning the language gives a sense of control, a feeling of mastery over technology that is finally not controllable but powerful beyond human comprehension. The longer I stayed, the more conversations I participated in, the less I was frightened of nuclear war.

How can learning to speak a language have such a powerful effect? One answer, discussed earlier, is that the language is abstract and sanitized, never giving access to the images of war. But there is more to it than that. The learning process itself removed me from the reality of nuclear war. My energy was focused on the challenge of decoding acronyms, learning new terms, developing competence in the language—not on the weapons and wars behind the words. By the time I was through, I had learned far more than an alternate, if abstract, set of words. The content of what I could talk about was monumentally different.

Consider the following descriptions, in each of which the subject is the aftermath of a nuclear attack:

Everything was black, had vanished into the black dust, was destroyed. Only the flames that were beginning to lick their way up had any color. From the dust that was like a fog, figures began to loom up, black, hairless, faceless. They screamed with voices that were no longer human. Their screams drowned out the groans rising everywhere from the rubble, groans that seemed to rise from the very earth itself.[4]

[You have to have ways to maintain communications in a] nuclear environment, a situation bound to include EMP blackout, brute force damage to systems, a heavy jamming environment, and so on.[5]

There is no way to describe the phenomena represented in the first with the language of the second. The passages differ not only in the vividness of their words, but in their content: the first describes the effects of a nuclear blast on human beings; the second describes the impact of a nuclear blast on technical systems designed to secure the 'command and control' of nuclear weapons. Both of these differences stem from the difference of perspective: the speaker in the first is a victim of nuclear weapons, the speaker in the second is a user. The speaker in the first is using words to try to name and contain the horror of human suffering all around her; the speaker in the second is using words to insure the possibility of launching the next nuclear attack.

Technostrategic language articulates only the perspective of the users of nuclear weapons, not the victims. Speaking the expert language not only offers distance, a feeling of control, and an alternative focus for one's energies; it also offers escape from thinking of oneself as a victim of nuclear war. No matter what one deeply knows or believes about the likelihood of nuclear war, and no matter what sort of terror or despair the knowledge of nuclear war's reality might inspire, the speakers of technostrategic language are allowed, even forced, to escape that awareness, to escape viewing nuclear war from the position of the victim, by virtue of their linguistic stance.

I suspect that much of the reduced anxiety about nuclear war commonly experienced by both new speakers of the language and long-time experts comes from characteristics of the language itself: the distance afforded by its abstraction, the sense of control afforded by mastering it, and the fact that its content and concerns are those of the users rather than the victims. In learning the language, one goes from being the passive, powerless victim to being the competent, wily, powerful purveyor of nuclear threats and nuclear explosive power. The enormous destructive effects of nuclear weapons systems become extensions of the self, rather than threats to it.

It did not take long to learn the language of nuclear war and much of the specialized information it contained. My focus quickly changed from mastering technical information and doctrinal arcana, to an attempt to understand more about how the dogma I was learning was rationalized. Since underlying rationales are rarely discussed in the everyday business of defence planning, I had to start asking more questions. At first, although I was tempted to use my newly acquired proficiency in technostrategic jargon, I vowed to speak English. What I found, however, was that no matter how well informed my questions were, no matter how complex an understanding they were based upon, if I was speaking English rather than expert jargon, the men responded to me as though I were ignorant or simple-minded, or both. A strong distaste for being patronized and a pragmatic streak made my experiment in English short-lived. I adopted the vocabulary, speaking of 'escalation dominance', 'pre-emptive strikes', and one of my favourites, 'sub-holocaust engagements'. This opened my way into long, elaborate discussions that taught me a lot about technostrategic reasoning and how to manipulate it.

But the better I became at this discourse, the more difficult it became to express my own ideas and values. While the language included things I had never been able to speak about before, it radically excluded others. To pick a bald example: the word 'peace' is not a part of this discourse. As close as one can come is 'strategic stability', a term that refers to a balance of numbers and types of weapons systems—not the political, social, economic, and psychological conditions that 'peace' implies. Moreover, to speak the word is to immediately brand oneself as a soft-headed activist instead of a professional to be taken seriously.

If I was unable to speak my concerns in this language, more disturbing still was that I also began to find it harder even to keep them in my own head. No matter how firm my commitment to staying aware of the bloody reality behind the words, over and over I found that I could not keep human lives as my reference point. I found I could go for days speaking about nuclear weapons, without once thinking about the people who would be incinerated by them.

It is tempting to attribute this problem to the words themselves—the abstractness, the euphemisms, the sanitized, friendly, sexy acronyms. Then one would only need to change the words: get the military planners to say 'mass murder' instead of 'collateral damage', and their thinking would change. The problem, however, is not simply that defence intellectuals use abstract terminology that removes them from the realities of which they speak. There *is* no reality behind the words. Or, rather, the 'reality' they speak of is itself a world of abstractions. Deterrence theory, and much of strategic doctrine, was invented to hold together abstractly, its validity judged by internal logic. These abstract systems were developed as a way to make it possible to, in Herman Kahn's phrase, 'think about the unthinkable—not as a way to describe or codify relations on the ground.

. . .

I was only able to make sense of this kind of thinking when I finally asked myself: Who—or what—is the subject? In technostrategic discourse, the reference point is not human beings but the weapons themselves. The aggressor ends up worse off than the aggressed because he has fewer weapons left; any other factors, such as what happened where the weapons landed, are irrelevant to the calculus of gain and loss.

The fact that the subjects of strategic paradigms are weapons has several important implications. First, and perhaps most critically, there is no real way to talk about human death or human societies when you are using a language designed to talk about weapons. Human death simply *is* collateral damage—collateral to the real subject, which is the weapons themselves.

Understanding this also helps explain what was at first so surprising to me: most people who do this work are on the whole nice, even good, men, many with liberal inclinations. While they often identity their motivations as being concern about humans, in their work they enter a language and paradigm that precludes people. Thus, the nature and outcome of their work can utterly contradict their genuine motives for doing it.

In addition, if weapons are the reference point, it becomes in some sense illegitimate to ask the paradigm to reflect human concerns. Questions that break through the numbing language of strategic analysis and raise issues in human terms can be easily dismissed. No one will claim that they are unimportant. But they are inexpert, unprofessional, irrelevant to the business at hand. The discourse among the experts remains hermetically sealed. One can talk about the weapons that are supposed to protect particular peoples and their way of life without actually asking if they are able to do it, or if they are the best way to do it, or whether they may even damage the entities they are supposedly protecting. These are separate questions.

This discourse has become virtually the only response to the question of how to achieve security that is recognized as legitimate. If the discussion of weapons was one competing voice in the discussion, or one that was integrated with others, the fact that the referents of strategic paradigms are only weapons might be of less note. But when we realize that the only language and expertise offered to those interested in pursuing peace refers to nothing but

weapons, its limits become staggering. And its entrapping qualities—the way it becomes so hard, once you adopt the language, to
stay connected to human concerns—become more comprehensible.

Within a few weeks, what had once been remarkable became unnoticeable. As I learned to speak, my perspective changed. I no longer
stood outside the impenetrable wall of technostrategic language
and once inside, I could no longer see it. . . .

My grasp on what I knew as reality seemed to slip. I might get
very excited, for example, about a new strategic justification for a
no-first-use policy and spend time discussing the ways in which its
implications for the US force structure in Western Europe were
superior to the older version. After a day or two I would suddenly
step back, aghast that I was so involved with the *military*
justifications for not using nuclear weapons—as though the moral
ones were not enough. What I was actually talking about—the mass
incineration of a nuclear attack—was no longer in my head.
. . .

Other recent entrants into this world have commented that while
the cold-blooded, abstract discussions are most striking at first,
within a short time you get past them and come to see that the language itself is not the problem.

I think it would be a mistake, however, to dismiss these early
impressions. While I believe that the language is not the whole
problem, it is a significant component and clue. What it reveals is a
whole series of culturally grounded and culturally acceptable mechanisms that make it possible to work in institutions that foster the
proliferation of nuclear weapons, to plan mass incinerations of millions of human beings for a living. Language that is abstract, sanitized, full of euphemisms; language that is sexy and fun to use;
paradigms whose referent is weapons; imagery that domesticates
and deflates the forces of mass destruction; imagery that reverses
sentient and nonsentient matter, that conflates birth and death,
destruction and creation—all of these are part of what makes it possible to be radically removed from the reality of what one is talking
about, and from the realities one is creating through the discourse.

Close attention to the language itself also reveals a tantalizing
basis on which to challenge the legitimacy of the defence intellectuals' dominance of the discourse on nuclear issues. When defence
intellectuals are criticized for the cold-blooded inhumanity of the
scenarios they plan, their response is to claim the high ground of

rationality. They portray those who are radically opposed to the nuclear *status quo* as irrational, unrealistic, too emotional—'idealistic activists'. But if the smooth, shiny surface of their discourse— its abstraction and technical jargon—appears at first to support these claims, a look below the surface does not. Instead we find strong currents of homoerotic excitement, heterosexual domination, the drive toward competence and mastery, the pleasures of membership in an élite and privileged group, of the ultimate importance and meaning of membership in the priesthood. How is it possible to point to the pursuers of these values, these experiences, as paragons of cool-headed objectivity?

While listening to the language reveals the mechanisms of distancing and denial and the emotional currents embodied in this emphatically male discourse, attention to the experience of learning the language reveals something about how thinking can become more abstract, more focused on parts disembedded from their context, more attentive to the survival of weapons than the survival of human beings.

. . .

When we outsiders assume that learning and speaking the language will give us a voice recognized as legitimate and will give us greater political influence, we assume that the language itself actually articulates the criteria and reasoning strategies upon which nuclear weapons development and deployment decisions are made. This is largely an illusion. I suggest that technostrategic discourse functions more as a gloss, as an ideological patina that hides the actual reasons these decisions are made. Rather than informing and shaping decisions, it far more often legitimizes political outcomes that have occurred for utterly different reasons. If this is true, it raises serious questions about the extent of the political returns we might get from using it, and whether they can ever balance out the potential problems and inherent costs.

I believe that those who seek a more just and peaceful world have a dual task before them—a deconstructive project and a reconstructive project that are intimately linked. Deconstruction requires close attention to, and the dismantling of, technostrategic discourse. The dominant voice of militarized masculinity and decontextualized rationality speaks so loudly in our culture that it will remain difficult for any other voices to be heard until that voice loses some of its power to define what we hear and how we name the world.

The reconstructive task is to create compelling alternative visions of possible futures, to recognize and develop alternative conceptions of rationality, to create rich and imaginative alternative voices—diverse voices whose conversations with each other will invent those futures.

Notes

1. General William Odom, 'C³I and Telecommunications at the Policy Level', incidental paper from a seminar, *Command, Control, Communications and Intelligence,* Spring 1980 (Cambridge, Mass., Harvard University Center for Information Policy Research) 5.
2. See Brian Easlea, *Fathering the Unthinkable: Masculinity, Scientists and the Nuclear Arms Race* (London: Pluto Press, 1983).
3. William L. Laurence, *Dawn Over Zero: The Study of the Atomic Bomb* (London: Museum Press, 1974), 198–9.
4. Hisako Matsubara, *Cranes at Dusk* (Garden City, New York: Dial Press, 1985).
5. General Robert Rosenberg, 'The Influence of Policy Making on C³I', speaking at the Harvard seminar, *Command, Control, Communications and Intelligence,* 59.

Part IV. Gender and Knowledge

12 The Mind's Eye

Evelyn Fox Keller and Christine R. Grontkowski

Feminist thought in the 1970s and 1980s echoes a number of themes familiar from radical thought of the 1960s. One such theme appears in the revolt against the traditional Western hierarchy of the senses. In this view, the emphasis accorded the visual in Western thought is not only symptomatic of the alienation of modern man, but is itself a major factor in the disruption of man's 'natural' relation to the world. The logic[1] of Western thought is too rooted in the visual; its failure, it is implied, derives from an unwholesome division of the senses.

Today, these themes appear in a new context and with a new specificity. There is a movement among a number of feminists to sharpen what, until now, had only been a vague sentiment weaving in and out of the major theme. The gist of this sentiment is that the logic of the visual is a male logic. According to one critic, what is absent from the logic which has dominated the West since the Greeks, and has been covered over by that logic, is woman's desire. 'Woman's desire', writes Luce Irigaray, 'does not speak the same language as man's desire. In this logic, the prevalence of the gaze . . . is particularly foreign to female eroticism. Woman find pleasure more in touch than in sight . . .'[2] In the same vein, Hélène Cixous dismisses Freudian and Lacanian theory of sexual difference for its 'strange emphasis on exteriority and the specular. A Voyeur's theory, of course.'[3]

The notion that vision is a peculiarly phallic sense, and touch a woman's sense, is, of course, not new. Indeed, it accords all too well with the belief in vision as a 'higher' and touch of 'lower' sense. As such, it has a long tradition, although not necessarily one that should be accepted. But the suspicion that the pervasive reliance on a visual metaphor marks Western philosophy as patriarchal is a

Reprinted with permission from Sandra Harding and Merrill Hintikka (eds.), *Discovering Reality* (Boston: D. Reidel, 1983), 207–24.

more general one. As such it needs to be explored. At the same time, however, such exploration needs to avoid the facile identifications of women with the 'lower' senses which are part of that same tradition.

This paper is written out of the conviction that, before these suspicions—feminist or otherwise—can be addressed explicitly, a thorough re-examination of the role which vision has played in Western thought needs to be undertaken. Vision is itself a complex phenomenon, with multiple subjective meanings. If it has had an historically phallic association, surely this association is rooted, not in biology, but in the cultural and psychological meanings attached to the visual. The following is an inquiry into the history of these meanings, entwined as they are in the role which the visual metaphor has played in Western epistemology. Only at the end do we consider, although briefly, the possible bearing that history has on the phallocentricity of our philosophical tradition. What we present is, therefore, primarily a philosophical and historical analysis of an issue which has important implications for feminist theory—implications which will need further elaboration elsewhere.

. . .

Hans Jonas has observed that, while Greek philosophy assumes the pre-eminence of vision among the senses from Plato on, 'neither he (Plato) nor any other of the Greek thinkers, in the brief treatments of sight which we have, seems to have really explained by what properties sight qualifies for these supreme philosophical honours'.[4] This may be because those properties are self-evident—certainly it has come to seem so to us. Yet it may not always have been so. In an elegant analysis of the transition from an oral tradition to a literate culture in ancient Greece, occuring between Homer and Plato, Eric Havelock has argued that not only has 'the eye supplanted the ear as the chief organ'[5] but that in the process a host of other changes was induced—changes from identification and engagement to individualization and disengagement, from mimesis to analysis, from the concrete to the abstract, from mythos to logos.[6] With the growing emphasis on the visual eye comes the growing development, even birth, Havelock argues, of the personal 'I'. It is perhaps no accident, then, that we first find a clear articulation of the pre-eminence of vision in Plato. But articulation does not mean argument, and it may be as important to note the general absence of Plato's direct argument on this subject as his indirect comments, which, by contrast, are amply present.

. . .

The particular pre-eminence which the visual enjoys is related, for Plato, to the pre-eminence light enjoys as a medium of perception, as well as to the pre-eminence which the sun enjoys among the divinities (or heavenly bodies). Of the special status of light he says: 'if light is a thing of value, the sense of sight and the power of being visible are linked together by a very precious bond, such as unites no other sense and its object.'[8] whereas of the special status and power of the sun he speaks constantly, albeit obliquely.

It is important here to consider how Plato's particular conception of vision affects its metaphoric uses. Vision is accomplished by a matching of like with like, first through a correspondence by likeness between the eyes and the sun ('the eye is the sense organ most similar to the sun' and 'The eye's power of sight is a kind of effusion dispensed to it by the sun.'[9]), and then through a matching of the various lights emanating from the eyes, the object, and the sun.

So when there is daylight around the visual stream, it falls on its like and coalesces with it, forming a single uniform body in the line of sight, along which the stream from within strikes the external object. Because the stream and daylight are similar, the whole so formed is homogeneous, and the motions caused by the stream coming into contact with an object or an object coming into contact with the stream penetrate right through the body and produce in the soul the sensation which we call sight.[10]

The mediation of perception (recognition) through likeness becomes a model for intellection as much as the eye itself becomes a model for the intellect. . . . All three components of the visual system—eye, the sun, and light—are used by Plato, both metaphorically and directly, to establish the characteristics of intelligibility.

The sun furnishes to visibles the power of visibility. . . . In like manner, . . . the objects of knowledge receive from the presence of the good their being known, but their very existence and essence is derived from it . . .[11]

As the good is in the intelligible region to reason and the objects of the reason, so is this (the sun) in the visible world to vision and the objects of vision.[12]

The correspondence between the visual and the mental interlocking of organ and object persists in Plato's conception of the origins of knowledge. This is perhaps most evident in his theory of *anamnesis*, recollection. In an attempt to explain the process of learning, Plato offers the following suggestion. The soul, before entering the body, once dwelt with the gods. There it enjoyed the same pure understanding of the cohesiveness of all things which was understood by the gods themselves. This knowledge was untrammeled by the senses; i.e. the senses were neither limitations to the grasp of real and true being, nor were they required as avenues of approach to it. Nevertheless, visual imagery is used for the description of the state of pure knowledge:

Every human soul has, by reason of her nature, had contemplation of true being . . . Beauty it was ours to see in those days when, amidst that happy company, we beheld with our eyes that blessed vision, our selves in the train of Zeus, others following some other god . . .; pure was the light that shone around us, and pure were we, without taint of that prison house which now we are encompassed, and call a body . . .[13]

Thus knowledge in its purest form is, for Plato, a state of being which is essentially divine. The soul as we experience it in this life is no longer unfettered but has to deal with the body which it inhabits. For living human beings the process of learning is recollection. The contact of our senses with some object or set of objects reminds us of the essential reality which we knew before.
. . .

Further, in the same breath with his affirmation of learning as recollection, Plato touches again on two of the other themes we have been stressing in this context: the affinity between seeing and knowing and the principle of kinship, or the meeting of like with like. This allows him to postulate the in-principle intelligibility of all things:

Thus the soul . . . since it has seen all things both here and in the other world, has learned everything that is. So we need not be surprised if it can recall the knowledge of . . . anything . . . it once possessed. All nature is akin, and the soul has learned everything, so that when a man has recalled a single piece of knowledge . . . there is no reason why he should not find out all the rest . . .[14]

The union, or reunion, of the soul with the Forms then constitutes knowing, just as the uniting of the light from the eye with the light from the sun constitutes seeing. Though that which mediates the meeting of the soul with the Forms is not specified, its analogy to light is often implicit. The terms which Plato uses for the Forms are *eidos* and *idea*, i.e. things which are seen.

The knower and that which is known, in this metaphor, are essentially kindred. They are both parts of the whole of being itself. Thus kinship with the universe and its structures constitutes Plato's metaphysical presupposition. His epistemological assumption is that we, who were originally part of the lawful divine structure, are thereby in principle able to see into (intuit) it fully again.

Modern science's confidence that nature, (properly objectified), is indeed knowable is surely derived from these Platonic concepts. Its confidence in the objectifiability of nature is, however, only partly derived from Plato. Two features of the scientific conception of objectifiability need to be distinguished. The first is the separation of subject from object, i.e. the distinction between the individual who perceives and the object which is perceived. The second is the move away from the conditions of perception, i.e. the separation of knowledge from the unreliability of the senses or, so to speak, the dematerialization of knowledge. The first is a move begun, as Havelock argues,[15] by Plato, but not completed until Descartes. The second is more thoroughly Platonic, and the greater part of two dialogues, the *Protagoras* and the *Theaetetus*, is devoted to the explication of the impossibility of basing knowledge on perception. In sum, he makes Theaetetus say: 'Taking it all together then, you call this perception . . . a thing which has no part in apprehending truth . . . nor, consequently, in knowledge either.'[16] It is precisely this aspect of the Platonic endeavour which makes the reliance on the visual metaphor for true knowledge most curious. We must ask whether there are not characteristics of vision, at least as conceived by Plato, which simultaneously invite the retreat from the body sought in Plato's epistemology and the maintenance of the moral–mystical character of his thought, in short, which constitute a paradox which pervades his work.

Indeed, there are several paradoxes implicit here. One is intrinsic to the conceptualization of knowledge as simultaneously objective and transcendent. The 'cool light of reason' establishes, in a single move, worldly distance and divine communion. Further, in allowing for the dissociation of truth from process, the visual metaphor ironically allows for the dissociation of the mental from the sensory. Vision is that sense which places the world at greatest remove; it is also that sense which is uniquely capable of functioning outside of time. It lends itself to a static conception of 'eternal truths'. Although itself one of the senses, by virtue of its apparent incorporeality, it is that sense which most readily promotes the illusion of disengagement and objectification. At the same time, it provides a compelling model for intangible communication offering the most profound and primitive satisfactions.

What appears to us so paradoxical in the different metaphoric functions served by vision and light must have been considerably less so to Plato, or for that matter, to all thinkers of the next two thousand years who accepted Plato's theory of vision. In Plato's understanding incorporeality and communion were not in conflict; the distinction between two kinds of looking or seeing which we have introduced had no place there—in fact, the very nature of the visual process he postulated incorporated both.[17] The light emitted by the eye was itself transcendent, and provided the very vehicle needed for the meeting of soul with soul. It is the 'stream from within' which, through its sympathetic coupling with the 'stream from without' can 'produce in the soul the sensation which we call sight'. As theories of vision underwent change, however, the different functions which the visual metaphor performed, and continued to perform, became considerably more paradoxical. With modern theories of optics, the eye becomes a passive lens, no longer thought to be emitting its own stream, and the transcendent coupling between inside and outside which Plato had imagined to occur was gone. With perception regarded as a passive recording, vision becomes a more suitable model for objectifiability and, at least ostensibly, a less adequate one for knowability. None the less, it continues as a model for both. How it does so constitutes something of a puzzle—a puzzle which, as we shall see, is somewhat resolved by the Cartesian split, but only by invoking a kind of 'seeing' other than that which the physicists had described.

By the seventeenth century these connections between light and knowledge were so much a part of the intellectual fabric that Descartes scarcely had to acknowledge his debt to Platonism in any of its forms. The analogy with the sun is no longer explicit—perhaps because the concept of light has become so closely affiliated with the process of intellection that no reference to a physical source is required. Vision and light, however, are frankly recognized by Descartes as analogous to the process of intellection. For example, in speaking of mental intuition, Descartes suggests that 'we shall learn how to employ our mental intuition from comparing it with the way in which we employ our eyes'.[18] And later, 'understanding apprehends by means of an inborn light' and inner perception must be perfected by the 'natural light' of reason. Ideas are reliable to the extent that they are 'clear', where

I term that 'clear' which is present and apparent to an attentive mind, in the same way that we see objects clearly when, being present to the regarding eye, they operate upon it with sufficient strength.[19]

Mind, for Descartes, is not only ontologically primary but a priori reliable; only the validity of aspects of our experience which are *not* purely mental requires explanation; the senses are repudiated *ab initio.* Descartes speaks of mental vision which is the means by which we know everything from the simplest to the most complex objects of knowledge.

those things which relatively to our understanding are called simple, are either purely intellectual or purely material . . . Those are purely intellectual which our understanding apprehends by means of a certain inborn light, and without the aid of any corporeal image.[20]

The inborn light receives its metaphysical dignity and stability by being totally and in a markedly Augustinian manner derived from divinity:

And so I very clearly recognize that the certainty and truth of all knowledge depends alone on the knowledge of the true God, in much that, before I knew Him, I could not have a perfect knowledge of any other thing. And now that I know Him I have the means of acquiring a perfect knowledge of an infinitude of things, not only of those which relate to God Himself, but also to those which pertain to corporeal nature . . .[21]

It is, in fact, with this kind of confidence that Descartes undertook the writing of his three treatises, the *Geometry*, the *Dioptrics*, and the *Meteors*, which he prefaced with the *Discourse on Method*. His purpose in the second of these texts was to examine in detail the existing science of optics, the theory of vision, and the actual workings of the human eye. The very first line of the *Dioptrics* reveals his attitude toward the importance of the faculty of sight: 'the whole of the conduct of our lives depends on our senses, among which that of sight is the most universal and the most noble . . .'[22]

His work on vision, perhaps even motivated by his commitment to both its literal and metaphoric importance, in fact led to an undermining of the suitability of sight as a metaphor for knowledge. Descartes's inquiries into the nature of vision and optics were of paramount importance in the Western acceptance of the copy theory of perception.[23] He, perhaps more than any other Western thinker, was responsible for laying the emission theory to rest,[24] with the result that the eye was henceforth regarded as a purely passive lens which simply receives the images projected upon it from without.

The consequences of this shift for our theories of knowledge were critical. It would seem, at this juncture, that we either accept the conclusion that knowledge itself is passive, or we abandon the visual metaphor. Not so, Descartes provided us with another alternative. He enabled us to retain *both* the conception of knowledge as active and the use of the visual metaphor by severing the connection between the 'seeing' of the intellect and physical seeing—by severing, finally, the mind from the body. He says, 'We know for certain that it is to the soul that that sense belongs, not to the body.'[25] And later, 'It is the soul that sees, not the eye'[26] although *not*, he tells us emphatically, with 'another pair of eyes'.[27]

To repeat then, we are arguing that our continuing reliance on the visual metaphor for knowledge inevitably implied that a change in our theories of one would induce changes in our theories of the other. In so far as it does not seem possible to conceive of knowledge as a passive recording of data—human pride alone would seem to preclude such an epistemological posture—then a sharper division between visual and mental sight was necessitated. It is this necessity which Descartes's dualistic philosophy provides a response to, which, we would argue, becomes a wedge for the mind–body dichotomy itself. As light and vision become more explicitly technical, physical phenomena, the eye itself a more mechanical device, the active knower is forced ever more sharply out of the bodily realm. The subject becomes finally severed from the objects of perception. With this move, the knowing agent has lost its last links to the percipient organism whose sense organs can now be relegated safely to the 'purely material'. Having made the *eye* purely passive, all intellectual activity is reserved to the 'I', which, however, is radically separate from the body which houses it.

The division of the world into mind and body has served to protect the active nature of an understanding which 'apprehends by means of an inborn light', of the 'natural light' of reason, from the passivity of a lens which merely records—in short, it has salvaged the possibility of knowledge. Nature may be visible by a mechanical process which leaves both the knower and the known disengaged, but it only becomes 'knowable' by virtue of an 'inborn' or 'natural' light which connects the mind's eye to truth. It is that light which re-establishes the subject's relation—a relation now totally and finally dematerialized—to the objects of perception.

The implications of this division for the concept of objectivity are traceable throughout the *Meditations*. *Meditation I* begins with the methodological doubt engendered by Descartes's suspicion of the subjectivity of the senses: their unreliability over the short term, their relationship to dreams and hallucinations, in short, of their epistemological uselessness. However, it is the conclusion of *Meditation VI*, that the material world really exists and can be known, which gives Descartes the scientific advantage. Having reached God through pure intellection, he can return, assured, to the material world which is now accessible through scientific, i.e.

objective reasoning. Specifically, experience is reliable to the extent to which it yields, in the final analysis, to the clear and distinct idea, i.e. to mathematics; thus objectified and measured, it is separated from the subject and can be evaluated accordingly.

The incorporation of these ideas into the scientific world view is evident throughout the subsequent history of science, but perhaps most notably in Newton's work. By the time of Newton, 'modern' theories of vision and optics were well ensconced. No longer can the visual mode provide the means of intermingling inside and outside which had been possible for Plato. More securely than ever, however, it can and does provide the means of establishing a total and radical severance of subject from object. The metaphor seems now to be cleansed of its 'coniunctio fantasies'. Communion with the Gods, or with Truth, must be established elsewhere. For Newton as for Descartes, pure thought, now emancipated from its visual dependence, is proposed to mediate this communion. The reliance on the senses with which Plato was ultimately and inextricably saddled, can finally be superseded, or so it would seem. Rational inquiry requires no sensory, physical intermediate to establish its one to one correspondence with the truths of nature. For example, Newton writes:

I do not define time, space, place and motion as being well known to all. Only I must observe that the vulgar conceive those quantities under no other notions but from the relation they bear to sensible objects. And thence arise certain prejudices, for the removing of which, it will be convenient to distinguish them into absolute and relative, true and apparent, mathematical and common.[28]

Nevertheless, both the eye and light remain of central importance to Newton, literally as well as metaphorically. In his methodology, psychology, metaphysics, even in his theology, the visual assumes throughout a position of prominence. Neither the thoroughgoing nor consistent rationalist that Descartes was, his strong commitment to empirical data is well known. Manuel has observed that 'Newton's eye was his first scientific instrument',[29] and so, in many ways, it remained. His first researches were indeed in optics, where his quasi-religious devotion to the measurement of optical diffraction patterns, performed to superhuman exhaustion and completion has been well documented.[30] These phenomena, as well as the many other optical phenomena he investigated, were studied with the naked eye. His early curiosity about colours led him to long bouts of direct gazing into the sun:

In a few hours I had brought my eyes to such a pass that I could look upon no bright object with either eye but I saw the sun before me, so that I durst neither write nor read to recover the use of my eyes shut myself up in my chamber made dark for three days together and used all means to divert my imagination from *the Sun. For if I thought upon him I presently saw his picture* though I was in the dark. But by keeping in the dark and employing my mind . . .[31]

His outrage, should any of his measurements be called into question, is also well known. Manuel comments: 'Scientific error was assimilated with sin, for it could only be the consequences of sloth on his part and a failure in his divine service.'[32]

But Newton's profound preoccupation with 'looking' seems to have had more to do with the inner eye than with the outer eye which he, as a physicist, knew well to be a mere lens—even though much of his 'looking' was done with the outer eye. Visual imagery is prominent in his remarks about theoretical inquiry as it is in his theory itself. 'When asked how he came to make his discoveries: "I keep the subject constantly before me, and wait till the first dawnings open slowly by little and little into the full and clear light."'[33] Indeed, as Heelan and others have noted, for Newton, 'The ideal of science was to "see" what God "saw",'[34] a belief illustrated so vividly in his many discussions of the Sensorium of God. In one passage, now famous, from the *Optics*, he wrote:

there is a Being incorporeal, living, intelligent, omnipresent who in infinite space, as it were in his sensory, sees things themselves intimately . . . of which things the images only (i.e. on the retina) carried through the organs of sense into our little sensoriums are there seen and beheld by that which in us perceives and thinks.[35]

Thus it seems evident that fantasies of union—union mediated by vision—retained a powerful hold on even so late a thinker as Newton. In spite of Newton's success in objectifying science, and in spite of the rejection of an interaction model of vision, we can see, in this quote, the residual influence of the Platonic metaphor. Communion through knowledge remains, even now, a central goal of science, and for both Newton and his successors, vision continues to provide an acceptable metaphoric model. To the extent that, for Newton, this communication could be achieved through both the physical and mental eye, vestiges of an earlier conception of vision are still evident. The subsequent history of science reveals a more thorough incorporation of the implications of modern theories of vision as the communicative functions of science are relegated more and more completely to thought.

Indeed, the history of science appears to have taken us on a long road of emancipation from the physical—somewhat paralleling Aristotle's hierarchy of the senses. Where Kepler experienced science as an opportunity to 'grasp (God), as it were, with my hand',[36] it was Newton's ambition to 'see' as God 'saw'. Einstein perhaps came closest to the ancient ideal when he concluded: 'I hold it true that pure thought can grasp reality.'[37] Throughout the history of scientific thought, then, the impact of the visual tradition continues to make itself felt, however residually. It is a tradition so deeply internalized by scientists that it no longer requires direct expression; indeed direct expression is perhaps no longer possible. Its indirect expression does, however, persist. The dual paradigm behind the promise of the visual—clarity and communion—survives as the root aspiration behind the dual tenets of modern science. In *objectifiability* the world is severed from the observer, illuminated as it were, by that sense which could operate, it was thought, without contaminating. In *knowability*, communion is re-established, mediated by a now-submerged but still evident dimension of the same sense. The persistence of theses tenets,[38] no longer quite appropriate, needs to be understood. One way of doing so is to seek to identify the philosophical moves which gave rise to them as we have attempted to do here. It would seem, from the present analysis, that our continuing commitment to the visual metaphor for knowledge is at least implicated in that persistence, and perhaps even in part responsible.

We have yet, however, to answer the basic question of why vision came to seem so apt a model for knowledge in the first place. A partial answer has been provided by Hans Jonas.[39] In an attempt to understand the characteristics of vision which are responsible for its particular appeal to classical philosophy, Jonas has conducted a phenomenological analysis of the senses. He finds three basic aspects of vision which provide grounds for its philosophical centrality. Under what he calls 'simultaneity of presence' he notes the distinctively spatial rather than temporal character of vision—a property uniquely responsible for our capacity to grasp the 'extended now'. He says:

Indeed only the simultaneity of sight, with its extended 'present' of enduring objects, allows the distinction between change and the unchanging and therefore between being and becoming. All the other senses operate by registering change and cannot make that distinction. Only sight therefore provides the sensual basis on which the mind may conceive the idea of the eternal, that which never changes and is always present.[40]

Under the heading of 'dynamic neutrality' he notes the peculiar lack of engagement entailed by seeing, the absence of intercourse. 'I see without the object's doing anything.'[41] 'I have nothing to do but to look, and the object is not affected by that: . . . and I am not affected.' to the 'neutralization of dynamic content' he attributes the expurgation of 'all traces of causal activity,' . . . 'The gain', he says, 'is the concept of objectivity'[42] but also, he notes earlier, the distinction between theory and practice. Indeed, he argues that it is through this very freedom effect by which the 'separation of contained appearance from intrusive reality gives rise to the separableness of *essentia* from *existentia* underlying the higher freedoms of theory'.[43] At the same time, however, it is precisely by virtue of its causal detachment that sight is the 'least "realistic" of the senses', and Jonas departs radically from Plato in concluding that when the 'underlying strata of experience, notably motility and touch' are rejected, 'sight becomes barren to truth'.[44]

Finally he notes a third dimension of vision which contributes critically to 'objectivity' and that is its uniquely advantageous dependence on distance. 'To get the proper view we take the proper distance,'[45] . . . 'and if this is great enough it can put the observed object outside the sphere of possible intercourse', and at the same time he observes that 'the facing across a distance discloses the distance itself as something I am free to traverse . . . The dynamics of perspective depth connects me with the projected terminus'[46] (even with infinity).

Jonas' analysis makes a number of important points . . . but neglects the ways in which vision as a model for knowledge can promote the sense of communion, of meeting of like with like, so central in Plato's understanding, which continues to survive in contemporary scientific belief. Though Jonas touches on an aspect of seeing which is connective, he fails to take account of an entire dimension of the visual experience not centrally contained in the experience of looking at, or surveying. That dimension is most dramatically captured in the experience of looking into, or 'locking eyes'—a form of communication and communion which is primitive and universally formative. In direct eye contact, we have a visual experience quite different from and in many ways even opposed to the sense of distance and objectivity evoked by merely looking at an object. The often highly charged experience of 'locking eyes' seems to do away with distance. As such, it may remain for all of us as a kind of paradigm for communication, for the connective aspects of vision.

In view of the preceding discussion, the question of a possible male bias to an epistemology modelled on vision has become considerably more complex than originally might have been thought. In particular, two facets of the metaphoric functions of vision need to be separated. The emphasis on the 'objectifying' function of vision, and the corresponding relegation of its communicative— one might even say erotic—function, needs to be separated from the reliance on vision as distinct from other sensory modalities. We suggest that if sexual bias has crept into this system, it is more likely to be found in the former than in the latter. Whatever germ of truth lies in Cixous' allegation of voyeurism may be more readily traced to the de-eroticization of the visual than to the traditional preoccupation with vision as such—a de-eroticization in fact promoted by classical theories of vision. Furthermore, in relegating the now submerged communicative aspects of the visual metaphor to the realm of thought, the latent eroticism in such experience is protected against by total disembodiment. The net result of such disembodiment is the same (as one of us has argued elsewhere[47]) as that implied in the radical division between subject and object assumed to be necessary for scientific knowledge: Once again, knowledge is safeguarded from desire. That the desire from which knowledge is so safeguarded is so intimately associated with the female (for social as well as psychological reasons) suggests an important impetus which our patriarchal culture provides for such disembodiment.[48] It is in this sense that Cixous is right.

Of course, the implications of ennobling vision above and beyond all the other senses also need to be examined, but we see no evidence to suspect in this an explicitly patriarchal move. Rather, it seems to express more diffuse cultural biases—biases which may, however, prove consonant with other more explicitly patriarchal biases. But even the last claim would be meaningless in the absence of alternative models. Our effort to articulate some of the influences the visual metaphor has had on our views of knowledge in general and scientific knowledge in particular would be futile if there were no other ways to describe knowledge. For us, the visual model seems almost inescapable. Yet, the question this paper must end with is whether or not that is so. Many authors have suggested that it is not a universal model, that it is not so prominent in other intellectual traditions. These suggestions lead us to ask: How might a conception of knowledge based on another metaphor differ? Some implications are immediately evident. Knowledge likened to the sense of hearing, for example, could not have made the same claims to atemporality, and might well lend itself more readily to a process view of reality. It is interesting to note in this regard that Heraclitus, our earliest temporal ontologist, evidently had a different metaphor in mind. In fact the verbal form of 'know' used by him, *ksuniemi*, originally meant 'to know by hearing'.[49] Similarly, a theory of knowledge which invokes the experience of touch as its base cannot aspire to either the incorporeality of the Platonic Forms, or the 'objectivity' of the modern scientific venture; at the very least it would have necessitated a more mediate ontology. We might agree with Vesey, who writes:

We can imagine a disembodied mind having visual experiences but not tactile ones. Sight does not require our being part of the material world in the way in which feeling by touching does. . . . The directness of seeing when contrasted with hearing, its non-involvement with the object when contrasted with feeling by touching, and its apparent temporal immediacy when contrasted with both feeling and hearing are features that may partly explain the belief that sight is the most excellent of the senses.[50]

But the crucial question which remains is whether it is possible to reconsider the criteria which lead to that conclusion. In a time when physics has once again altered our conception of vision and light, when we know that neither the apparent atemporality nor the 'dynamic neutrality' of vision are features of reality, but only of our relatively coarse daily observations it seems appropriate to reassess our commitment to the ideals which these features imply.

Notes

1. The use of the term 'logic' here is colloquial; it refers to the general structure of thought rather than to a system of formal language.
2. Luce Irigaray, 'This Sex Which Is Not One', in E. Marks and I. de Courtrivon (eds.), *The New French Feminisms* (Amherst: University of Massachusetts Press, 1980), 101.
3. Hélène Cixous, 'Sorties', *The New French Feminisms*, 92.
4. Hans Jonas, 'The Nobility of Sight', *Philosophy and Phenomenological Research*, 14 (1954), 507.
5. Eric A. Havelock, *Preface to Plato* (Universal Library, 1967).
6. For development of the same theme in different terms, see Bruno Snell, *The Discovery of the Mind*, trans. T. G. Rosenmeyer (Cambridge, Mass.: Harvard University Press, 1953), esp. ch. 9.
7. *Timaeus*, 61d–68c.
8. Ibid. 508a.

9. Ibid. 508b.
10. *Timaeus*, 45c–d.
11. *Republic*, 509b.
12. Ibid. 508c.
13. *Phaedrus*, 250a–c.
14. Ibid. 81c–d.
15. He attributes to Plato in particular the 'self imposed task' of establishing 'two main postulates: that of the personality which thinks and knows, and that of a body of knowledge which is thought about and known'.
16. *Theaetetus*, 186e.
17. For an analysis of the Greek vocabulary of seeing, cf. Snell, ch. 1.
18. René Descartes, *Regulae* IX from E. S. Haldane and G. R. T. Ross (eds.), *Philosophical Works of Descartes* (New York: Dover, 1934), i.
19. *Principles*, i. 45–6, Haldane and Ross (eds.), 237.
20. *Regulae XII*, Haldane and Ross (eds.), i. 41.
21. *Meditation V*, Haldane and Ross (eds.), i. 183.
22. Our translation from *Descartes, Oeuvres et Lettres* (Paris: Bibliothèque de la Pleiade, 1953), 181.
23. For a discussion of a much earlier articulation of the copy theory of perception, see Vasco Ronchi, *Optics: The Science of Vision*, trans. Edward Rosen (New York, 1957).
24. Some recent critics have argued for a residual presence of the emission theory in Descartes's work. See, for example, Stephen H. Daniel, 'The Nature of Light in Descartes' Physics', *The Philosophic Forum*, 7 (1976), 341, but this point is debatable and does not in any case mitigate against his essential and explicit rejection of optical emanations.
25. *Dioptrics in Descartes: Philosophical Writings* trans. and ed. Elizabeth Anscombe and Peter T. Geach (London, 1954), 242.
26. Ibid. 253.
27. Ibid. 246.
28. Isaac Newton, *Principia*, quoted by Alexandre Koyré in *From the Closed World to the Infinite Universe* (New York, 1958), 161.
29. Frank Manuel, *A Portrait of Newton* (Cambridge, Mass.: Harvard University Press, 1968), 78.
30. Ibid. 78–9.
31. Ibid. 85.
32. Ibid. 141.
33. Ibid. 86.
34. Patrick A. Heelan, 'Horizon, Objectivity and Reality in the Physical Sciences', *International Philosophical Quarterly*, 7 (1967), 407. In this context cf. also E. A. Burtt, *The Metaphysical Foundations of Modern Science*, ch. 7 (New York, 1932).
35. *Opticks*, 3rd edn. (London, 1721), 344.
36. Gerald Holton, *Thematic Origins of Scientific Thought* (Cambridge, Mass.: Harvard University Press, 1973), 86.
37. Ibid. 234.
38. See, e.g. E. F. Keller, 'Cognitive Repression in Contemporary Physics', where ongoing dispute over the interpretation of quanum mechanics is traced to the retention of one or the other of these dual tenets. Some psychological sources of the appeal to belief in the 'objectifiability' and the 'knowability' of nature are suggested.

39. Jonas, 'The Nobility of Sight'.
40. Ibid. 513.
41. Ibid. 514.
42. Ibid. 515.
43. Ibid. 514.
44. Ibid. 517.
45. Ibid. 518.
46. Ibid.
47. Evelyn Fox Keller, 'Gender and Science', reprinted in this volume.
48. See n. 38 for further discussion of this point, as well as E. F. Keller, 'Nature as "Her"', *Proceedings of the Second Sex Conference* (New York University, 1979).
49. Cf., Kurt Von Fritz, '*Noûs, Noein* and Their Derivation in Pre-Socratic Philosophy', in Alexander P. D. Mourelatos (ed.), *The Pre-Socratics* (New York, 1974).
50. G. N. A. Vesey, 'Vision', *The Encyclopedia of Philosophy* (New York, 1967), viii. 252.

13 Though This Be Method, Yet There Is Madness in It: Paranoia and Liberal Epistemology

Naomi Scheman

When you do not see plurality in the very structure of a theory, what do you see?

María Lugones, 'On the Logic of Pluralist Feminism'

Somewhere every culture has an imaginary zone for what it excludes, and it is that zone we must try to remember today.

Catherine Clément, *The Newly Born Woman*

In an article entitled 'The Politics of Epistemology', Morton White argues that it is not in general possible to ascribe a unique political character to a theory of knowledge.[1] In particular, he explores what he takes to be the irony that the epistemologies developed by John Locke and John Stuart Mill for explicitly progressive and democratic ends have loopholes that allow for undemocratic interpretation and application. The loopholes White identifies concern in each case the methods by which authority is granted or recognized.

Neither Locke nor Mill acknowledges any higher epistemic authority than human reason, which they take (however differently they define it) as generic to the human species and not the possession of some favoured few. But for both of them, as for most other democratically minded philosophers (White discusses also John Dewey and Charles Sanders Peirce), there needs to be some way of distinguishing between the exercise of reason and the workings of something else, variously characterized as degeneracy, madness, immaturity, backwardness, ignorance, passion, prejudice, or some other state of mind that permanently or temporarily impairs the development or proper use of reason. That is, democracy is seen as needing to be defended against 'the excesses of unbridled relativism and subjectivism' (White 1989: 90).

© 1993 by Westview Press. Reprinted with permission from Louise Antony and Charlotte Witt (eds.), *A Mind of One's Own* (Boulder, Colo.: Westview Press, 1993).

The success of such a defence depends on the assumption that if we eliminate the voices of those lacking in the proper use of reason, we will be eliminating (or at least substantially 'bridling') relativism. This, I take it, can only mean that those whose voices are listened to will (substantially) agree, at least about those things that are thought to be matters of knowledge, whether they be scientific or commonsense statements of fact or fundamental moral and political principles or specific judgements of right or wrong. To some extent this assumption is tautological: it is frequently by 'disagreeing' about things the rest of us take for granted that one is counted as mad, ignorant, or otherwise not possessed of reason. But precisely that tautologousness is at the root of what White identifies as the loophole through which the antidemocratic can pass: moral, political, and epistemological élitism is most attractive (to the élite) and most objectionable (to others) when the non-élite would say something different from what gets said on their behalf, allegedly in the name of their own more enlightened selves.

. . .

A striking feature of the advance of liberal political and epistemological theory and practice over the past three hundred years has been the increase in the ranks of the politically and epistemically enfranchised. It would seem, that is, that the loopholes have been successively narrowed, that fewer and fewer are being relegated to the hinterlands of incompetence or unreliability. In one sense, of course, this is true. Race, sex, and property ownership are no longer explicit requirements for voting, officeholding, or access to education in most countries. But just as exclusionary gestures can operate to separate groups of people, so similar gestures can operate intrapsychically to separate those aspects of people that, if acknowledged, would disqualify them from full enfranchisement. We can understand the advance of liberalism as the progressive internalization—through regimes of socialization and pedagogy—of norms of self-constitution that (oxymoronically) 'democratize privilege'.

Thus various civil-rights agendas in the United States have proceeded by promulgating the idea that underneath the superficial differences of skin colour, genitalia, or behaviour in the bedroom, Blacks, women, and gays and lesbians are really just like straight white men. Not, of course, the other way around: difference and similarity are only apparently symmetrical terms. In the logic of political identity, to be among the privileged is to be among the same, and for the different to join those ranks has demanded the willingness to separate the difference-bearing aspects of their identity, to demonstrate what increasingly liberal regimes were increasingly willing to acknowledge: that one didn't need, for example, be a man to embrace the deep structure of misogyny. It is one of my aims to argue that the norms that have structured modern epistemic authority have required the internalization of such exclusionary gestures, the splitting off and denial of (or control over) aspects of the self that have been associated with the lives of the disenfranchised, and that those gestures exhibit the logic of paranoia.

. . .

I am also concerned with the extent to which much work is still captive to older pictures, notably in the continuing dominance of individualism in the philosophy of psychology. A fully social conception of knowledge that embraces diversity among knowers requires a corresponding conception of persons as irreducibly diverse and essentially interconnected. The individualism of modern personhood entails a denial both of connection and of individuality: modern subjects are distinct but not distinctive. Philosophers have taken this subject as theirs: it is his (*sic*) problems that have defined the field, the problems of anyone who takes on the tasks of internalizing the norms of privilege. As these norms change, so must the corresponding conceptions of personhood.

It is in this light that I want to examine the influence of Descartes's writings, works of intentionally democratic epistemology that explicitly include women in the scope of those they enfranchise. I have argued elsewhere, as have many others,[2] for the undemocratic nature of the influence of Cartesian epistemology, an influence that extends even to those epistemologies standardly treated as most antithetical to it (notably, empiricism). In particular, I want to argue that the structures of characteristically modern epistemic authority (with science as the central paradigm) normalized strategies of self-constitution drawn from Cartesian Method. The discipline that is meant to ensure that proper use of the Method will not lead to 'unbridled relativism and subjectivism', although intended by Descartes to be both liberatory and democratic, has come to mirror the repressions that mark the achievement of privilege. Those strategies find, I believe, a peculiarly revelatory echo in the autobiographical writings of Daniel Paul Schreber and in their use in Freud's theory of paranoia.[3] Ironically, by the very moves that were meant to ensure universal enfranchisement, the epistemology that has grounded modern science and liberal politics not only has provided the means for excluding, for most of its history, most of the human race but also has constructed, for those it authorizes, a normative paranoia.

DESCARTES

> They are, in essence, captives of a peculiar arrogance, the arrogance of not knowing that they do not know what it is that they do not know, yet they speak as if they know what all of us need to know.[4]

Cartesian philosophy is a paradigmatic example of White's thesis about the subversion of the democratic intent of an epistemology, although not because of Descartes's own views about who it authorized. Descartes's explicit intent was the epistemic authorization of individuals as such—not as occupiers of particular social locations, including the social location of gender.[5] Most important, Descartes wanted to secure epistemic authority for individual knowers, who would depend on their own resources and not on the imprimatur of those in high places, and, he argues, those resources could only be those of mathematized reason, not those of the senses. Only such a use of reason could ensure the sort of stability that distinguishes knowledge from mere opinion. Descartes's Method was designed to allow anyone who used it to place him- or herself beyond the influence of anything that could induce error. Human beings, he argues, were not created as naturally and inevitably subject to error: God wouldn't have done that. What we are is finite, hence neither omniscient nor infallible. But if we recognize our limits and shield ourselves from the influence of what we cannot control, we can be assured that what we take ourselves to know is, in fact, true.

The Method is a form of discipline requiring acts of will to patrol a perimeter around our minds, allowing it only what can be determined to be trustworthy and controlling the influence of the vicissitudes of our bodies and of other people. Purged of bad influences, we will be struck by the 'clarity and distinctness' of truths like the cogito. We will have no real choice but to acknowledge their truth, but we ought not to find in such lack of choice any diminution of our freedom. Because the perception of truth comes from within us, not 'determined by any external force', we are free in assenting to it, just as we are free when we choose what we fully and unambivalently want, even though it makes no sense to imagine that, given our desire, we might just as well have chosen otherwise.[6]

Freedom from determination by any external force requires, for Descartes, freedom from determination by the body, which is, with respect to the mind, an external force. Thus when Descartes invokes the malicious demon at the end of the First Meditation to help steel him against lazily slipping back into credulity,[7] his efforts are of a piece with his presentation at the end of *The Passions of the Soul* of 'a general remedy against the passions'.[8] Passions are no more to be dispensed with entirely than are perceptions (or, strictly speaking, *other* perceptions, given that passions are for Descartes a species of perception). But no more than other perceptions are passions to be taken at face value: They can be deceptive and misleading. Still less are they to be taken uncritically as motives to act, whether the action in question be running in fear from the dagger I perceive before me or assenting to its real existence. In both cases, I (my mind) need to exercise control over my perceptions or, at least, over what I choose to do in the face of them. Seeing ought *not* to be believing in the case of literal, embodied vision, but when ideas are seen by the light of reason in the mind's eye, assent does and should follow freely.

The individualism of Cartesian epistemology is yoked to its universalism: though we are each to pursue knowledge on our own, freed from the influence of any other people, what we come up with is not supposed to be our own view of the world—it is supposed to be the truth, unique and invariable. When Descartes extols, in the *Discourse*,[9] the greater perfection of buildings or whole towns that are the work of a single planner over those that sprang up in an unco-ordinated way, he may seem to be extolling the virtues of individuality. But what he finds pleasing are not the signs of individual style; it is the determining influence of reason as opposed to chance. Individualism is the route not to the idiosyncrasies of individuality but to the universality of reason.

This consequence is hardly accidental. Scepticism, which was a tool for Descartes, was for some of his contemporaries the ultimate, inevitable consequence of ceding epistemic authority to individual reason. If epistemic democratization was not to lead to the nihilism of the Pyrrhonists or the modesty of Montaigne, Descartes needed to demonstrate that what his Method produced was knowledge, not a cacophony of opinion.[10] It could not turn out to be the case that the world appeared quite different when viewed by people differently placed in it. More precisely, everyone had to be persuaded that if it *did* appear different from where they stood, the remedy was to move to the Archimedean point defined by the discipline of Cartesian Method. Those who could not so move were, in the manner of White's discussion, relegated to the ranks of the epistemically disenfranchised.

. . .

It is central to Descartes's project, as it is to the social and political significance of that project, that no one and nothing other than agents themselves can confer or confirm epistemic authority (despite God's being its ultimate guarantor: His guarantee consists precisely in our each individually possessing such authority). Epistemic authority resides in the exercise of will that disciplines one's acts of assent—principally to refrain from assenting to whatever is not perceived clearly and distinctly.[11] And the will, for Descartes, is not only equally distributed among all people but is also, in each of us, literally infinite: what is required is not the acquisition of some capacity the exercise of which might be thought to be unequally available to all; rather, it is the curbing of a too-ready willingness to believe.

. . .

But, as I argued above, there is no reason why philosophers' own views about who can and cannot fully exemplify their requirements of epistemic enfranchisement should carry any special weight when the question concerns the democratic or anti-democratic effect of their theories, especially as those theories have been influential far beyond those philosophers' lifetimes. Descartes is a paradigmatic case in point.

The Cartesian subject was revolutionary. The individual bearer of modern epistemic authority became, through variations on the originating theme of self-constitution, the bourgeois bearer of rights, the self-made capitalist, the citizen of the nation-state, and the Protestant bound by conscience and a personal relationship to God. In Descartes's writings we find the lineaments of the construction of that new subject, and we see the centrality of discipline to its constitution. Such discipline is supposed in theory to be available to all, not only to those whose birth gave them a privileged place in the world. If one was placed where one could not see the truth, or obtain riches, or exercise political or religious freedom, the solution was to move to some more privileged and privileging place. The 'New World' was precisely constituted by the self-defining gestures of those who moved there from Europe and who subsequently got to determine who among those who followed would be allowed to take a stand on the common ground. (That constitution of the 'New World' is one reason why the people who already lived there merited so little consideration in the eyes of those who invaded their home. The relationship the Indians took—and take—themselves to have to the land, a relationship grounded in their unchosen, unquestionable ties to it, was precisely the wrong relationship from the perspective of those who came to that land in order to define themselves anew by wilfully claiming it, unfettered by history.)

With the success of the revolutions prefigured in the Cartesian texts, it became clear that the theoretical universalism that was their underpinning existed in problematic tension with actual oppression. Those who succeeded in embodying the ideals of subjecthood oppressed those whose places in the world (from which, for various reasons, they could not move[12]) were (often) to perform the labour on which the existence and well-being of the enfranchised depended and (always) to represent the aspects of embodied humanness that the more privileged denied in themselves.
. . .

Privilege, as it has historically belonged to propertied, heterosexual, able-bodied, white men, and as it has been claimed in liberal terms by those who are variously different, has rested on the successful disciplining of one's mind and its relation to one's body and to the bodies and minds of others. The discourses of gender, race, class, and physical and cognitive abilities have set up dichotomies that, in each case, have normalized one side as the essentially human and stigmatized the other, usually in terms that stress the need for control and the inability of the stigmatized to control themselves. Acts of violence directed against oppressed groups typically are presented by their oppressors as pre-emptive strikes, justified by the dangers posed by the supposedly less-civilized, less-disciplined natures of those being suppressed. Work-place surveillance through lie detectors and drug testing (procedures in which subjects' bodies are made to testify to the inadequacies of their minds and wills), programmes of social control to police the sexual behaviour of homosexuals, the paternalistic disempowerment of the disabled, increasing levels of verbal and physical attacks on students of colour by other students, and the pervasive terrorism of random violence against women all bespeak the need on the part of the privileged to control the bodies and behaviour of those who are 'different', a need that both in its targets and in its gratuitous fierceness goes beyond securing the advantages of exploitation.
. . .

Despite Descartes's genuinely democratic intentions, as his epistemology was taken up by those who followed him, it authorized those—and only those—whose subject positions were constituted equally by their relationship to a purportedly objective world and by their relationship to the disenfranchised Others, defined by their inescapable, undisciplined bodies.

PARANOIA, DISCIPLINE, AND MODERNITY

. . .

As Freud argues, the central mechanism of paranoia is projection, that process by which something that had been recognized as a part of the self is detached from it (a process called 'splitting') and reattached onto something or someone other than the self. An underlying motivation for such splitting is narcissism: what is split off is incompatible with the developing ego. But it is significant to note that one obvious effect is the diminution of the self—it no longer contains something it once did. One consequence of that recognition is that it provides a motivation for thinking of that which is split off as wholly bad, perhaps even worse than it was thought to be when it was first split off. It has to be clear that the self really is better off without it.

This is one way of thinking about the fate of the body in Cartesian and post-Cartesian epistemology. The self of the cogito establishes its claim to authority precisely by its separation from the body, a separation that is simultaneously liberating and totally isolating. Although Descartes goes on, under the protection of God, to reclaim his body and to place himself in intimate and friendly relation to it, the loss to the self remains: René Descartes, along with all those who would follow his Method, really is a *res cogitans*, not a sensual, bodily person.

. . .

Such a self, privileged by its estrangement from its own body, from the 'external' world, and from other people, will, in a culture that defines such estrangements as normal, express the paranoia of such a stance not only through oppression but, more benignly, through the problems that are taken as the most fundamental, even if not the most practically pressing: the problems of philosophy. Those problems—notably, the mind–body problem, problems of reference and truth, the problem of other minds, and scepticism about knowledge of the external world—all concern the subject's ability or inability to connect with the split-off parts of itself—its physicality, its sociability. Such problems are literally and unsurprisingly unsolvable so long as the subject's very identity is constituted by those estrangements. A subject whose authority is defined by his location on one side of a gulf cannot authoritatively theorize that gulf away. Philosophers' problems are the neuroses of privilege; discipline makes the difference between such problems and the psychosis of full-blown paranoia.

BEYOND MADNESS AND METHOD

> The new mestiza copes by developing a tolerance for ambiguity. . . . She has a plural personality, she operates in a pluralistic mode—nothing is thrust out, the good, the bad and the ugly, nothing rejected, nothing abandoned.[13]

> The alternative to relativism is partial, locatable, critical knowledges sustaining the possibility of webs of connections called solidarity in politics and shared conversations in epistemology.[14]

The authorized subject thus achieves and maintains his authority by his ability to keep his body and the rest of the world radically separated from his ego, marked off from it by policed boundaries. Within those boundaries, the self is supposed to be unitary and seamless, characterized by the doxastic virtue of noncontradiction and the moral virtue of integrity. The social mechanisms of privilege aid in the achievement of those virtues by facilitating splitting and projection: the unity of the privileged self is maintained by the dumping out of the self—on to the object world or on to the different, the stigmatized Others—everything that would disturb its pristine wholeness.

Various contemporary theorists are articulating alternative conceptions of subjectivity, conceptions that start from plurality and diversity, not just among but, crucially, within subjects.[15] From that starting point flow radically transformed relationships between subjects and between subjects and the world they would know.

One way to approach these discussions is to return to Freud. Mental health for Freud consisted in part in the acknowledgement by the ego of the impulses of the id: 'where id was, there ego shall be.'[16] the German is more striking than the English: the German words for 'ego' and 'id' are 'ich' and 'es';[17] the sense is 'where *it* was, there *I* shall be.' One can take this in two ways. Under the sorts of disciplinary regimes that constitute epistemic privilege, the exhortation has a colonizing ring to it. The not-I needs to be brought under the civilizing control of the ego; the aim is not to split it off but to tame it. Splitting represents the failure of colonization, the loss of will for the task of domestication. The healthy ego is unified not because it has cast out parts of itself, but because it has effectively administered even the formerly unruly outposts of its dominion. Or so goes the story one is supposed to tell. (Any splitting goes unacknowledged.)

There is another way to take Freud's exhortation. The aim might be not to colonize the 'it' but to break down the distinction between 'it' and 'I', between object and subject. 'Where it was, there I shall be', not because I am colonizing it, but because where I am is always shifting. As Nancy Chodorow puts it, in giving an object-relational alternative to the classical Freudian account, 'where fragmented internal objects were, there shall harmoniously related objects be'.[18] Moving becomes not the installment of oneself astride the Archimedean point, the self-made man taming the frontier of the 'New World', but the sort of 'world' travel María Lugones discusses as the ground of what she calls, following Marilyn Frye, 'loving perception'.[19] By putting ourselves in settings in which we are perceived as—and hence are able (or unable not) to be—different people from who we are at home, we learn about ourselves, each other, and the world. And part of what we learn is that the unity of the self is an illusion of privilege, as when, to use Lugones's example (from a talk she gave at the University of Minnesota), we think there is a natural, unmediated connection between intention, will, and action, because if we are privileged, the world collaborates with us, making it all work, apparently seamlessly, and giving us the credit. As Frye puts it, we are trained not to notice the stagehands, all those whose labour enables the play to proceed smoothly.[20]

What is problematic about Descartes's Faustian gesture is not the idea that the world is in some sense our creation. Rather, it is on the one hand the individualism of the construction (or, what comes to the same thing, the unitary construction by all and only those who count as the same, the not-different) and on the other the need to deny any construction, to maintain the mutual independence of the self and the world. Realism ought not to require such independence on the side of the world, any more than rationality ought to require it on the side of the knowing subject, if by realism we mean the recognition that the world may not be the way anyone (or any group, however powerful) thinks it is and if by rationality we mean ways of learning and teaching that are reliably useful in collective endeavours.

Philosophical realism has typically stressed the independence of the world from those who would know it, a formulation that, at least since Kant, has been linked with the intractability of scepticism. But it's hard to see exactly why independence should be what is required. A world that exists in complex interdependence with those who know it (who are, of course, also part of it) is none the less real. Lots of real things are not independent of what we think about them, without being just what anyone or any group takes them to be—the economy, to take just one obvious example. The interdependencies are real, as are the entities and structures shaped by them. One way we know they are real is precisely that they look different to those differently placed in relation to them. (There aren't a variety of diverse takes on my hallucinations.) The only way to take diversity of perspectives seriously is to be robustly realistic, both about the world viewed and about the material locations of those doing the viewing. Archimedean, difference-denying epistemology ought to be seen as incompatible with such a robust realism: how could there possibly be one account of a world shaped in interaction with subjects so diversely constituted and with such diverse interests in constructing and knowing it?

A specifically Cartesian feature of the conception of the world as independent is the world as inanimate, and consequently not reciprocally engaged in the activities through which it comes to be known. Thus, for example, the social sciences, which take as their objects bearers of subjectivity and the entities and structures they create, have been seen as scientifically deficient precisely because of the insufficiently independent status of what they study. (The remedy for such deficiency has typically been the dehumanizing objectification of the 'subjects' of the social sciences, an objectification especially damaging when those subjects have been otherwise oppressed.) But it's far from obvious that being inert should make something more knowable. Why not take 'subject' and 'object' to name not ontological categories but reciprocal, shifting positions? Why not think of knowledge emerging paradigmatically in mutual interaction, so that what puzzles us is how to account not for the objectivity of the social sciences but for the intersubjectivity of the natural sciences?[21]

In a discussion of the problems from an African-American perspective, with the critical legal theorists' rejection of rights, Patricia
Williams suggests that rather than discarding rights,

society must give them away. Unlock them from reification by giving them
to slaves. Give them to trees. Give them to cows. Give them to history.
Give them to rivers and rocks. Give to all of society's objects and untouch-
ables the rights of privacy, integrity, and self-assertion; give them distance
and respect. Flood them with the animating spirit that rights mythology
fires in this country's most oppressed psyches, and wash away the shrouds
of inanimate-object status.[22]

One might respond similarly to the suggestion from postmodernist
quarters that we discard subjectivity and agency; rather, we should
profligately give them away and invest the things of the world with
subjectivity, with the ability and interest to return our gaze.[23]
Realism can mean that we see ourselves as inhabiting a world in
which the likes of us (whoever we may be) are not the only sources
of meaning, that we see ourselves as implicated in, reciprocated by,
the world.

The world as real is the world as precisely not dead or mechanis-
tic; the world as trickster, as protean, is always slipping out from
under our best attempts to pin it down.[24] The real world is not the
world of our best physics but the world that defeats any physics that
would be final, that would desire to be the last word, 'the end of the
story, the horizon of interpretation, the end of "the puzzlement"', a
desire Paul Smith calls 'claustrophilic'.[25] Donna Haraway imagina-
tively sketches an epistemology for the explicitly partial, fragmentary, ununified knowers we are and need to be if we are to move
within and learn from the complexities of the world and the complexities of how we are constructed in it. As she puts it, 'Splitting,
not being, is the privileged image for feminist epistemologies of scientific knowledge' (Haraway 1988: 586).
. . .

The unity of privileged subjectivity is mirrored in the demand
that language be transparent, a demand most explicit in the now-
discredited ideal languages of the logical positivists but lingering in
the demands of present-day analytic philosophers for (a certain picture of) clarity, as though the point of language was to be seen
through. When June Jordan writes of Black English that one of its
hallmarks is 'clarity: if the sentence is not clear it's not Black
English', she might seem to be endorsing such a demand, but the
clarity she extols is contextual and 'person-centered'. 'If your idea,
your sentence, assumes the presence of at least two living and active

6. Cartesian philosophy was, in fact, influential on and in some ways empowering for contemporary feminists. See Ruth Perry (1985), 'Radical Doubt and the Liberation of Women', *Eighteenth-Century Studies*, 18, 472–93.

7. Descartes, Meditation 4, in *The Philosophical Writings of Descartes* (1985), 2 vols., tr. John Cottingham and Dugald Murdoch (Cambridge: Cambridge University Press), 40 (hereafter C&M).

8. Meditation 1, C&M, 2.15.

9. Descartes, *The Passions of the Soul*, pt. III, sec. 211, C&M, 1.403. I owe the suggestion to look again at *The Passions of the Soul* to Adam Morton, who may, however, have had something else entirely in mind.

10. Descartes, *Discourse on the Method*, pt. II, C&M, 1.116–17.

11. On Pyrrhonist and Montaignean scepticism, see Richard Popkin (1960), *The History of Scepticism from Erasmus to Descartes* (New York: Humanities Press).

12. See Margaret Dauler Wilson (1978), *Descartes* (London: Routledge and Kegan Paul), 17–31.

13. The most heinous case of such oppression is slavery, and the US slave trade, of course, required the movement of slaves from their homes. But such movement was the denial rather than the expression of those people's will, and it served to confirm what, in the non-literal sense, was their place in the world as defined by Europeans and Euro-Americans, part of which was that they had no say over where, literally, their place in the world was to be.

14. Gloria Anzaldúa (1987), *Borderlands/La Frontera: The New Mestiza* (San Francisco: Spinsters/Aunt Lute), 79.

15. Donna Haraway (1988), 'Situated Knowledges: The Science Question in Feminism and the Privilege of Partial Perspective', *Feminist Studies*, 14/3 (Fall), 584.

16. Sandra Harding and Donna Haraway are two such theorists. See also three papers in which María Lugones develops a pluralistic theory of identity: (1982) 'Playfulness, "World"-Traveling, and Loving Perception', *Hypatia*, 2/2 (Summer), 3–19; (1990) 'Hispaneando y Lesbiando: On Sarah Hoagland's *Lesbian Ethics*', *Hypatia*, 5/3 (Fall), 138–46; and (1991) 'On the Logic of Pluralist Feminism', in Claudia Card (ed.), *Feminist Ethics* (Lawrence: University Press of Kansas), 35–44.

17. Freud, *New Introductory Lectures on Psychoanalysis*, SE, 22.80.

18. See Bruno Bettelheim (1983), *Freud and Man's Soul* (New York: A. A. Knopf). The *New Introductory Lectures* were originally written in English, but the point still holds: Freud used the English of his translators.

19. Nancy Chodorow (1986), 'Toward a Relational Individualism: The Mediation of Self Through Psychoanalysis', in Thomas C. Heller, Morton Sosna, and David E. Wellbey (eds.), *Reconstructing Individualism: Autonomy, Individuality, and the Self in Western Thought* (Stanford, Calif.: Stanford University Press), 197–207.

20. Lugones, 'Playfulness, "World"-Traveling, and Loving Perception'; Marilyn Frye (1983), 'In and Out of Harm's Way: Arrogance and Love', in *Politics of Reality: Essays in Feminist Theory* (Freedom, Calif.: Crossing Press), 52–83.

21. Frye, 'To Be and Be Seen', *Politics of Reality*, 167–73.

22. For a start on such an account, as well as an argument for why we should seek one, see Lorraine Code (1991), *What Can She Know? Feminist Theory and the Construction of Knowledge* (Ithaca, NY: Cornell University Press), esp. chs. 3 and 4; and Sandra Harding (1991), *Whose Science? Whose Knowledge?*

people, you will make it understandable because the motivation behind every sentence is the wish to say something real to somebody real.'[26] The clarity of analytic philosophy, by contrast, is best exhibited in argumentative contexts, detached from the specificities of anyone's voice, in avoidance of *ad hominem* and other genetic fallacies. The clarity of Black English, Jordan explains, is grounded in the rhythms and intonations of speech, in the immediacy of the present indicative, and in an abhorrence of abstraction and the eschewal of the passive (non)voice. It is the clarity of illumination, not of the transparent medium. In contrast to the language of philosophy, which assumes its adequacy as a vessel for fully translatable meaning, Black English does not take its authority for granted. It is a language 'constructed by people constantly needing to insist that we exist, that we are present'.[27] It aims not at transparent representation but as subversive transformation; it is an act of intervention, used by communities of resistance and used within those communities for collective self-constitution.

. . .

María Lugones recounts her experiences as a 'multiplicitous being', a US Latina lesbian who could not be unitary without killing off a crucial part of who she is, without betraying both herself and others with whom she identifies and for whom she cares.[28] Without identification with and engagement in struggle within *la cultura hispana Nuevomejicana*, the imperilled community in which she 'has found her grounding', she risks becoming 'culturally obsolete', but as a lesbian within that culture, she is not a lover of women— she is an 'abomination'. Needing to be both of the very different people she is in the *Nuevomejicana* and lesbian cultures, she works not for unity but for connection, for the not-to-be-taken-for-granted understanding of each of her selves by the other, understanding that is cultivated by work in the 'borderlands', 'the understanding of liminals'. Victoria Davion contends that it is such connection that can ground a conception of integrity that does justice—as she argues any usable feminist notion of integrity must— to the experiences of multiplicitous beings.[29]

. . .

I want to suggest that, without blurring the specificities of such experiences, we can recognize that the experiences even of those who identify with dominant cultures can lead, in different ways, to multiplicitous identities. Gloria Anzaldúa, for example, stresses the importance for *mestizas* of the acceptance of all of who they are, 'the white parts, the male parts, the queer parts, the vulnerable parts'.[30]

But she equally calls for such self-acceptance on the part of the privileged, as the only alternative to the splitting and projection that underwrite domination: 'Admit that Mexico is your double, that we are irrevocably tied to her. Gringo, accept the doppelganger in your psyche. By taking back your collective shadow the intracultural split will heal' (Anzaldúa 1987: 86).

Erica Sherover-Marcuse suggests that all children are subject to what she calls 'adultism', a form of mistreatment which targets all young people who are born into an oppressive society.[31] Such mistreatment, she argues, is 'the "training ground" for other forms of oppression', a crucial part of the socialization of some as oppressor, some as oppressed, and most of us into complex combinations of both. Central to such socialization is its normalization, the denial of its traumatic nature, the forgetting of the pain; and central to emancipation is 'a labor of *affective remembrance*'.[32] Alice Miller argues similarly in *For Your Own Good* that only those who have been abused become abusers, and her account focuses on the mechanisms of splitting and projection: 'Children who have grown up being assailed for qualities the parents hate in themselves can hardly wait to assign those qualities to someone else so they can once again regard themselves as good, "moral", noble, and altruistic' (Miller 1984: 91).

The abuse of which Alice Miller writes, which ranges from the normative to the horrific, shares the requirement of amnesia, which means that the split-off parts of the self, whether they be the survival-ensuring 'alters' of the multiple or the stigmatized Others of the privileged, are empathically inaccessible. What Sherover-Marcuse calls 'an emancipatory practice of subjectivity' (Sherover-Marcuse 1986: 140) requires memory, connection, and the learning of respect for the Others that we are and for the Others outside of us. . . . As long as we hold on to the ideal of the self as a seamless unity, we will not only be marginalizing the experiences of those like María Lugones and Gloria Anzaldúa, for whom such unity could only be bought at the price of self-betrayal, but we will be fundamentally misrepresenting the experiences of even the most privileged among us, whose apparent unity was bought at the price of the projection onto stigmatized Others of the split-off parts of themselves that they were taught to despise.

As Quine has persuasively argued,[33] epistemology cannot come from thin air: to naturalize epistemology is to acknowledge that we need to study how actual people actually know. But one thing we ought to know about actual people is that they inhabit a world of systematic inequality, in which authority—centrally including epistemic authority—is systematically given to some and withheld from others. If our interest is in changing that world, we need to look critically at the terms of epistemic authority. Certainly there is no reason why those who have historically been dominated by the epistemology of modernity—the objects to its subjects—should accept the terms of that epistemology as the only route to empowerment.

That epistemology presents itself as universal, a universal defined by precisely that which is not different in the ways that some are defined as different: women (not men), people of colour (not white people), the disabled (not the able-bodied), gays and lesbians (not heterosexuals). To again echo Foucault, none of these categories is natural or ahistorical, and they all came into existence as strategies of regimentation and containment. They all represent aspects of the multiple, shifting, unstable ways that people can be, aspects that have been split off from the psyches of the privileged, projected onto the bodies of others, and concretized as identities. The privileged, in turn, having shucked off what would threaten their sense of control, theorize their own subjectivity (which they name generically human) as unitary and transparent to consciousness and characterized by integrity and consistency. Not only is such subjectivity a myth; its logic is that of paranoia.

Notes

1. Morton White (1989), 'The Politics of Epistemology', *Ethics*, 100 (Oct.) 77–92.
2. See Genevieve Lloyd (1984), *The Man of Reason* (Minneapolis: University of Minnesota Press); Susan Bordo (1984), *The Flight of Objectivity: Essays on Cartesianism and Culture* (Albany: SUNY Press); my (1987) 'Othello's Doubt/Desdemona's Death: The Engendering of Scepticism', in Judith Genova (ed.), *Power, Gender, Values* (Edmonton, Alberta: Academic Printing and Publishing); and Jacquelyn Zita (1989), 'Transsexualized Origins: Reflections on Descartes's *Mediations*', *Genders*, 5 (Summer), 86–105.
3. Daniel Paul Schreber (1988), *Memoirs of My Nervous Illness*, tr. and ed. Ida Macalpine and Richard A. Hunter (Cambridge, Mass.: Harvard University Press); Sigmund Freud (1958), 'Psycho-Analytic Notes upon an Autobiographical Account of a Case of Paranoia (Dementia Paranoides)', *Standard Edition* (hereafter *SE*), 12:9–82 (London: Hogarth Press). [Editors' note: The author's extended discussion of the Schreber case-history has been deleted for reasons of space. Readers should consult sections I and III of the original unabridged essay for details.]
4. Alice Miller (1984), *For Your Own Good: Hidden Cruelty in Child-Rearing and the Roots of Violence*, tr. Hildegarde Hannum and Hunter Hannum (New York: Farrar, Straus, and Giroux), 91.
5. Molefi Kete Asante (1987), *The Afrocentric Idea* (Philadelphia: Temple University Press), 4.

Thinking from Women's Lives (Ithaca, NY: Cornell University Press), esp. ch. 4.

23. Patricia J. Williams (1991), *The Alchemy of Race and Rights: Diary of a Law Professor* (Cambridge, Mass.: Harvard University Press), 165.
24. See Rainer Maria Rilke's 'Archaic Torso of Apollo'; 'There is no place / that does not see you. You must change your life' (1938) *Translations from the Poetry of Rainer Maria Rilke*, trans. M. D. Herter Norton (New York: W. W. Norton).
25. Haraway, 'Situated Knowledges', 596.
26. Paul Smith (1988), *Discerning the Subject* (Minneapolis: University of Minnesota Press), 98.
27. June Jordan, (1985) 'Nobody Mean More to Me than You/And the Future Life of Willie Jordan', in *On Call: Political Essays* (Boston: South End Press, 1985), 129–30. Such accounts make evident the Eurocentrism of deconstructive sorties against such notions as presence, voice, and authorship. See, for example, Jacques Derrida (1981), 'Plato's Pharmacy', in his *Dissemination*, trans. Barbara Johnson (Chicago: University of Chicago Press).
28. Jordan, 'Nobody Mean More to Me than You', 129.
29. Lugones, 'Hispaneando y Lesbiando'.
30. Victoria M. Davion, 'Integrity and Radical Change', in Card (ed.), *Feminist Ethics*, 180–92.
31. Anzaldúa, *Borderlands/La Frontera*, 88.
32. Erica Sherover-Marcuse (1986), *Emancipation and Consciousness: Dogmatic and Dialectical Perspectives in the Early Marx* (Oxford: Blackwell), 139.
33. Ibid. 140. Emphasis in original.
34. W. V. Quine (1969), 'Epistemology Naturalized', *Ontological Relativity and Other Essays* (New York: Columbia University Press).

14 A Science of Mars or of Venus?

Mary Tiles

> Western science, from Heraclitus to Hiroshima, has only
> known martial Nature . . . The laws of Venus–Mother Nature
> cannot be deciphered by the children of Mars . . .[1]

For as long as there has been anything worthy of the name of science, there have been those who have criticized its claim to superior
knowledge. With the birth and prodigious growth of modern science, the corresponding growth of critical opinion led, in the eighteenth century, to a divorce of the sciences from the humanities
around which our educational institutions, and our universities in
particular, have been built. It is this divorce which renders problematic the status of the social or human sciences. For the extent to
which man can be an object of scientific knowledge will be questioned by those insisting on an opposition between human knowledge and values as embodied in the humanities, and the
dehumanized objective knowledge proclaimed within the natural
sciences.

It is not my purpose here to become embroiled in any of these
long-standing debates. They are mentioned merely to contrast criticisms of the natural sciences which are based on a presumption of
this dichotomy with recent, internal critiques of natural science.
These latter critiques come from a number of rather diverse sources
(feminism, the movements for alternative medicine and appropriate technology, advocates of time irreversibility in physics, some
African philosophers and others claiming to speak for the Third
World, for example). There is, however, a common theme—the
need to extend the horizons of science by breaking down the rigid
division between the natural and the human sciences *not* by sub-

Reprinted from *Philosophy*, 62 (July 1987). By permission of Cambridge
University Press.

suming the human sciences under the natural sciences as presently constituted, but by reconsidering the project of the natural sciences themselves. In short the call is for development of science with a human face, science which aims more at co-operating with than at conquering Nature, which learns more by conversing or conducting a dialogue with Nature than by putting it on the rack, to force it to reveal its secrets, a science of Venus rather than a science of Mars.

Such criticisms are potentially more radical in their effect than those of any anti-science movement, for the latter presume the *status quo*, accept the battle lines as currently drawn up and are prepared for combat which both sides know can never be mortal. The former, however, seek neither to reject nor to mutilate science but to enrich it by changing it. However, the two types of criticism have not always been distinguished, for they do have one element in common—they both rest on a rejection of values claimed to lie at the heart of modern, technological science. The customary defence against such criticism is to deny that values play any significant role in the natural sciences. If successful such a move can insulate science against criticism without there being any need for further discrimination amongst its critics. Thus we see increasing numbers of philosophers of science rallying under the banner of Realism to defend the view of science as aiming at objective Truth and as possessed of methods of theory choice which, even if they do not guarantee truth, do at least ensure objectivity by preventing the intrusion of non-scientific interests or values into theory choice.

I shall argue, however, that this move will not serve as a defence against the possibility of radical, internal critiques of modern science for two reasons. First, because there are crucial unclarities in its proposed 'ultimate criterion'[2] of theory choice, unclarities which serve to mask the intrusion of those very values to which critics most frequently take exception. Secondly because it is a defence mounted in the wrong place; it is a defence of science as constituted by the content of its theories, whereas the attack falls on science as delimited and determined by its methods and on the attitudes which lie behind their use in the normative definition of science and the scientific. In other words a defence of theory is mounted by appeal to method, but the attack falls on aspects of the implementation of that very method and on its formative role in science, rather than falling directly on theoretical content:

This 'logic of domination' is not merely a problem of the abuse of science or its misapplication; it is rather embedded in the methodology itself. By this method, means and ends have become completely scrambled.[3]

221

POSITIONS

Isaiah Berlin[4] describes the modern scientific tradition as the tradition of those who believe (i) that it is possible to make steady progress in the entire sphere of human knowledge, (ii) that methods and goals are, or should be, ultimately identical throughout this sphere, and (iii) that we have reached a stage where the achievements of the natural sciences are such that it is possible to derive their structure from a single set of clear principles which, if correctly applied, make possible indefinite further progress in unravelling the mysteries of Nature. He goes on to link this with a tradition in Western thought which he traces back to Plato and characterizes as resting on the following assumptions:

(a) Every genuine question has exactly one true answer; all others being false.
(b) The method which leads to correct solutions to all genuine problems is rational in character and is, in essence, if not in detailed application, identical in all fields.
(c) The solutions, whether or not they are discovered, are true universally, eternally, and immutably: true for all times, places, and men.

. . .

The voices urging departure from this tradition reject one or more of assumptions (a)–(c) and thus also the practical implication drawn from them. For example:

Those development professionals who claim to speak for the Third World stress the way in which the priorities of the affluent societies within which scientific experts are educated (sophisticated armaments, diseases of the overfed and ageing, multiplicities of costly drugs, high-input mechanized agriculture, and so on) distract scientific attention away from the problems facing poorer countries to the extent that these problems are not visible to the expert:

The values and preferences of first professionals are typically polar opposites of last realities . . . Most professionals see first values as sophisticated and scientific, and last realities as primitive and based on ignorance.[5]

Marxist historians of science have linked the rise to dominance of scientific/technological modes of thought with the rise of capitalism and hence with class struggle.

. . .

Feminists, on the other hand, have seen scientific culture as the product of a distinctively male consciousness, embodying forms of rationality which are alien to women. They have found in this an explanation for the continued underrepresentation of women in science. But as Evelyn Fox Keller argues,[6] the rejection of science as 'male' by women is a temptation which should be resisted. To follow this path is to leave rationality and objectivity in the male domain, securely dominated by men, thus forcing women to retreat to a purely 'female' subjectivity.

. . .

She claims that there are, and always have been, elements in scientific culture which embody values traditionally classified as 'female'. The opposition between conceptions of science as 'dominating' and as 'conversing with' nature is seen as representing

. . . a dual theme played out in the work of all scientists, in all ages. But the two poles of this dialectic do not appear with equal weight in the history of science. What we therefore need to attend to is the evolutionary process that selects one theme as dominant . . . (op. cit. 600).

Her suggestion is that it is in the process of selection that ideological forces come into play and that it is therefore at this level that masculine ideology will be found to have influenced perceptions of the history of science and hence also the conception of science itself.

A recent voice speaking for one of Fox Keller's less dominant themes has been that of Ilya Prigogine.[7] He sees the basic conflicting attitudes of the 'two cultures' as grounded in an opposition between the atemporal view of classical science and the historical time-orientated view of the social sciences and humanities. His claim (based on his work in thermodynamics) is that modern science is rediscovering time (op. cit. n. 7, p. xxviii) and that to the extent that time-irreversible processes become fundamental in physics, science itself will have to abandon its atemporal vantage point. One source of conflict between scientific and non-scientific world views would then have been removed:

Must we choose between a science that leads to alienation and an antiscientific metaphysical view of nature? We think such a choice is no longer necessary, since the changes that science is undergoing today lead to a radically new situation (op. cit. n. 7, p. 7).

. . .

THE DEFENCE OF SCIENTIFIC NEUTRALITY

The first move made in defence of the view that science is value-free, using objective standards to select theories which approximate ever more closely to the truth about the physical world, is concessionary. It has to be admitted that political and economic concerns frequently influence the direction of science. Similarly it has to be admitted that the predominance of white middle-class males from developed countries in the scientific professions may affect the selection of research topics. In other words it is allowed that non-scientific interests and values have a role in determining the *direction* of scientific research. What is rejected is the claim that values have a role in determining the *content* of science. Scientific results, if acceptable at all, should be acceptable to anyone, regardless of his concerns, because the ultimate standard of acceptability is empirical adequacy. It is ultimately Nature, not the scientist, who determines which theory will survive.

It will then be suggested that all potentially radical critiques of science in fact fall on those factors which determine the direction of scientific research, and not on factors determining the results of that research once its direction is determined. It is admitted that these factors may currently be such as to encourage research in those areas which hold promise of military application, to discourage research on matters (such as development of a male contraceptive pill) which might be expected to be of more concern to women than men, to discourage medical research on, for example, cheap preventive measures or the effectiveness of homoeopathic remedies, which might not be in the interests of the medical establishment, to discourage research on low-input agriculture, which is not in the interests of the chemical companies funding agricultural research, and so on. But the fault here lies with political and social formations, with the professional organization and funding of science, not with science or its methods as such. Military, male-dominated, first-world science is, and can be, no different *qua science* from pacific, female-dominated or Third World science.

In order to complete this argument it is necessary to insist on the distinction between the practice of science and its cognitive content. For science is being defended only as a putative body of objective knowledge; the totality of current scientific practice is not being defended. However, this distinction cannot be made as sharply as seems to be supposed. For the defence of science as a body of knowledge is made by appeal to method. It is argued that in the natural sciences empirical adequacy provides an ultimate objective criterion.[8] Given a choice between two competing theories, that which is empirically more successful is that which should be chosen. Where relative merits on this score are unclear, experimental testing of both should be pursued. There may be situations in which theory choice is not *determined* by relative empirical adequacy, but there can be no going against this standard. It cannot be overridden by other concerns, resulting in choice of the less empirically adequate theory. Globally it is argued that the manifest success of science at the level of conferring increased ability to predict and control suggests that non-scientific interests have not in fact played any significant role in theory choice (empirical adequacy has not been overridden). Hence such interests have not played any significant role in determining the cognitive content of science.

There are two claims here. The first is a *normative* claim about how theory choice *should* go. The second is an *evaluative* claim that science has been successful. This success is itself taken as evidence that past choices have in fact (by and large) been made as good scientific choices should have been made (in such a way as to exclude non-scientific interests and values).

It has been argued[9] that this prescription and this evaluation are grounded in an interest which is universal. Every human being in his right mind (every rational person) has an interest in increasing his technological control over his environment, for only in this way can his living conditions be improved and his life be made more secure. If this is the case, then increased ability to predict and control should command universal respect. But from the fact that in some sense of the words everyone values the ability to predict and control, it does not follow that everyone must or should respect and exclusively value the particular forms of technological control achieved by modern science. This is a point to which we shall return after completing the strategy of the Realist defence.

The Realist tends to focus attention on scientific laws and theories (the content of science), insisting that both the truth-or-falsity and scientific acceptance or rejection of any proposed law is independent of whatever factors may have conditioned the route to its proposal. The context of discovery should not be confused with the context of justification. The strength of the claim that values play no role in determining the content of science derives largely from the plausibility of the claim that laws such as the second law of thermodynamics, or of gravitational attraction, could not, and would not, be rejected by any of those mounting critiques of modern science. They do not propose, even as possible, societies equipped with perpetual motion machines, air transport provided by courtesy of lumps of matter not affected by gravitational forces, and in which kettles are boiled by standing them on icebergs. It is because critics of modern science do not reject the laws of its core theories (interpreted as acceptance of their truth), that it is argued that science (its cognitive content) is independent of interests and values.

BREACHING THE DEFENCES: THE POSSIBILITY OF RADICAL CRITIQUE

But this is to miss the point of the more sophisticated critiques of science, for what is under attack is precisely exclusive attention on this conception of cognitive content. What is attacked is the idea that any law or theory which withstands tests devised in a certain way could ever represent the whole truth on any matter. The claim that science, including its content, is conditioned by interests and values, is not the claim that content is *determined* by these conditions, but that it is *limited* by them. This does not require rejection of scientific results. The mistake made in failing to recognize the conditioned nature of the cognitive claims of science is that of proclaiming as the whole truth (truth *simpliciter*) that which is, and can only ever be, a partial truth, something which is not to be rejected, but which is not to be over-extended either. Newtonian mechanics, still highly useful, must be distinguished from Newtonian mechanism as a world view, which no longer plays the role of guiding scientific research. Similarly scientific laws in general must be distinguished from the world view determined by projecting them along with the methods of modern science as the whole truth and as sole arbiters of truth. A conditioned view is not false, but partial. What is to be rejected is the claim that it is unconditioned. The two-valued rhetoric of truth and falsity must be avoided if there is to be room for non-destructive criticism.

Now it is not at all clear that one can rest with the idea that the cognitive content of science is simply a sum of claims made in individual statements in its theories. An integral component of the conception of *scientific*, as opposed to general, knowledge is that it is theoretical in the sense of being systematically organized. Laws do not merely function as true or false statements but serve in the provision of explanations and in the organization and classification of phenomena. Theories impose a structure, and this structure is arguably part of the cognitive content of the theory. To claim that a theory is correct is not just to claim the truth of its laws taken individually but also to make claims about the interrelations between laws and hence also between the phenomena to which they apply. Some laws, such as the conservation of energy, are treated as being more fundamental than others, such as Hooke's law, and fundamental laws of a given theory (such as the laws of motion in Newton's mechanics) are unlikely to be independent parts of that theory. Within theories laws are given an epistemological value based on their theoretical, explanatory role rather than on their mere empirical adequacy.

. . .

If, then, the content of science is extended to cover theories and not merely laws, it would appear that place must be accorded at least to the notion of *epistemological value*. The epistemological value accorded to a law or group of laws may change even though they continue to be widely used and accepted. But in this case their significance, their scientific, theoretical content will have changed. Criticisms which bear on epistemological values can thus have a bearing on the content of science, and epistemological values are themselves a function not only of empirical adequacy but also of explanatory ideals and conceptions of the goal of scientific endeavour.

This can be further illustrated by returning to the argument which goes from the manifest success of science to the absence of influence by non-scientific interests. When we ask for evidence of the success of science, we find ourselves directed to the marvels of modern technology. This suggests, following the pattern of inference used, that past theory choices have been made with an eye to technological applications. But we might well ask whether an interest in technology is an intrinsically scientific interest. Was this what was meant by the criterion of empirical adequacy? What interests are internal to science? This depends on how the goal of science is characterized.

The new science proclaimed by Bacon and prefigured in the new astrology and magic of the Renaissance magi seeks to shift the *primary* focus of scientific attention away from contemplatively perceived truth to the goal of mastery over nature. The pursuit of truth is no longer disinterested; the interest is in increasing man's ability to dominate and control. Knowledge is sought and valued to the extent that it confers this ability.

. . .

It is thus quite correct to say, with Newton-Smith,[10] that without an interest in prediction and control modern science would not exist. But it should at the same time be noted (a) that this interest is an expression of a particular conception of the nature of Man and of his relation to Nature, (b) how this interest determines the conception of what constitutes scientific knowledge—the knowledge sought, and (c) how it also determines the empirical, evidential base of science.

(a) The scientific, intellectual interest in prediction and control presupposes a view of Man on which such an interest is worthy of his dignity, an expression and fulfilment of a distinctively human potential, which raises man above the animals and the rest of nature.

(b) The knowledge which confers power is primarily knowledge of laws of action and of the contexts in which they are applicable. It dictates a shift away from the Aristotelian goal of knowledge of natures qualitatively expressed in terms of natural dispositions and tendencies. This is a classic illustration of a shift of epistemological values, for the knowledge most valued by Aristotelians is not rejected by Baconians, it is indeed still necessary, but it is no longer the goal. Similarly, knowledge of regularities was recognized by Aristotelians, but did not form the goal of Aristotelian science; those disciplines dealing in regularities (astronomy, optics) were not granted the status of true sciences.

(c) Knowledge of how to divert the course of nature cannot be obtained merely by observing the natural course of events, by watching nature take her course. For the aim is not merely to imitate nature but to dominate it, to achieve things never naturally achieved (how else are supernatural powers to be revealed?). Thus the primary learning will be from attempts to manipulate, from artefacts and from artificially contrived situations. Here too there is a shift of epistemological values. The observation of artefacts available to Aristotelians was discounted by them as not productive of the sort of knowledge required by natural philosophers. Bacon turns the tables, placing much less value on naturally occurring phenomena and much more on experimental situations. Internal critiques of science seek to redress the balance, rescuing knowledge of the natural from its seventeenth-century devaluation.

Here there is a very strong sense in which interest in mastery over nature shapes the content of science; it determines what form interesting knowledge shall take. It involves redrawing the conception of the natural thought of as that which does not require explanation. It can no longer be that which occurs naturally, in the course of nature, or that which is explicable by reference to the nature of a thing. It is now that which occurs according to a basic law of action, when this is not subject to interference. The whole conception of what needs and what does not need explanation, of what is fundamental and what derivative, is changed.

But if the aim of science is to obtain knowledge which confers power, there may be several distinct clarifications of this aim, depending on the further specification of the power sought. Evidence of the success of science, evidence that the knowledge presented does indeed confer power requires exercise of that power. If the power sought is that of technological innovation, then production of new technology is evidence of success. If the power sought is merely the ability to dominate and conquer Nature, then obliteration of pests, development of powerful defoliants and other weapons of destruction is evidence of success. If the power sought is power to improve the lot of Mankind in general, and of its poorer groups in particular, then ability to provide on a large scale food, water, housing in stable communities viable not just for a limited period, but in the long term, and without sacrificing the opportunity for others to have the same benefits, would count as success.

The Davy lamp was heralded as a success for science. Davy had solved a problem, that of producing a lamp which would burn in a methane-rich atmosphere without causing an explosion. But in what sense was this a success for science? Davy had analysed the problem, come up with a theoretical solution and tested it in his laboratory. But as a miner's safety lamp it was hardly a success. In the conditions in which it was used down the mines, it did cause explosions. Davy lacked the knowledge of mining necessary to perceive the difference between his problem, which he had solved, and the problem of producing a lamp which had the required property when used down a mine, which he had not solved. The problem put before Davy was the mine-owner's problem of how to get a light that would function in the methane-rich atmosphere of 'crept' workings. The miners' concern, on the other hand, was the reduction of accident rates. The Davy lamp benefited the mine owners, but at the expense of the miners. (The lamp was introduced in 1816.

In Durham and Northumberland in 1798–1816 there were twenty-seven explosions and 447 deaths, in 1817–35 there were forty-two explosions and 538 deaths.) Was this then a success to be notched up for science?[11]

What has to be recognized is that ·the knowledge which is sufficient for success in the laboratory, or for observational success, does not on its own confer the other powers, nor can the methods for achieving this success automatically be assumed to be those which will yield success in wider spheres. The question of the extent to which this knowledge is necessary to and has been productive of success in wider fields is the central issue in debates about the relation between science and technology and about the role of technology in the solution of human problems. The theoretical and explanatory endeavour of the natural sciences has been analytic and foundational—the quest for fundamental laws, for basic constituents of matter, the basic building blocks by reference to which the characteristics of all more complex objects and phenomena are to be explained. The empirical investigative methods which form part of this endeavour have been correspondingly analytic, concentrating on the isolation of causal factors, on the creation of controlled experimental conditions in which the effects of variations in single variables can be studied. But the application of knowledge gained in this way depends on skills and knowledge of a different kind. Analysis has to be accompanied by synthesis, work in controlled environments has to be extended to work in much more variable conditions. Knowledge of the variable conditions of application is as important as knowledge of fundamental theory, practical skill is as important as theoretical understanding.

It is easy to see how emphasis on the analytic approach can lead away from any immediate practical concerns to total immersion in fundamental problems, whose answers raise further problems of their own. Puzzle-solving is fun in itself and does not have to be related to any further pay-off. Disinterested science as a puzzle-solving activity with a momentum of its own is quite understandable. What then becomes problematic is the relation of this autonomous activity and its results to wider concerns.

The fate of Scholastic philosophy should serve here both as an illustration and as a warning. The fact that modern science is experimental does not insure it against scholasticism (immersion in a puzzle-solving tradition divorced from its roots and hence ultimately direction-less and of little relevance). Laboratory experiments involve manufactured conditions and problems raised by these conditions can spawn one another just as easily as pure theoretical problems. When empirical success means success purely relative to laboratory experiments the standard becomes wholly internal to the research tradition and prediction and control is narrowed to ability to predict what will happen in carefully controlled conditions together with the ability to produce those conditions. The valuation of the theoretical over the practical tends to restrict the content of science to the point where its practical value may be called into question.

These debates cannot be pursued further here. The present point is simply that once the notion of scientific interest is extended beyond truth to some form of power or control, the question of what constitutes a scientific and what a non-scientific interest is no longer clear cut. This is not to reject the extension, for it is only this move which makes empirically grounded, theoretical knowledge of the world possible. The defenders of scientific neutrality are right to stress the importance of empirical success as a criterion for the assessment of claims to scientific knowledge, if wrong to think that it follows that science is value-free, that is no room for debate about what constitutes success, or that the nature of the scientific project is beyond dispute. Absence of values is not a precondition of objectivity.

Even if it is maintained that science aims solely at knowledge, at the construction of true theories about the physical world so that its only internal interest is in truth and in empirical adequacy as a necessary condition of truth, empirical adequacy can still mean *either* predictive observational success, constituted by agreement with a predefined, fixed class of possible observations of natural occurrences (saving the phenomena), or experimental success, the prediction and successful production of novel, artificial, phenomena. The difference between these clearly constitutes a difference in the evidential base against which scientific theories are assessed. The significance of the difference can be brought out by considering the change which occurs with the shift to modern, experimental science, which is, as we have seen, dictated by a changed conception of the sort of knowledge sought. This change was in turn seen to be related to the value placed on that knowledge and to the wider interests that it was hoped would be served by obtaining it.

There are two components present at the dawn of modern science[12] and it is important to see that acceptance of one does not entail acceptance of the other. First there is a reorientation of values in which active involvement with the physical world and a concern for material well-being come to be seen as expressive of human dignity, rather than as something to be despised by the philosopher, the seeker after truth. The disinterested contemplation of truth was previously valued for its effective distancing of man from the material world. It was a disengagement from the technological and the practical and thus expressive of human dignity defined more strictly in terms of spiritual and intellectual, as opposed to material, life. But of course the reorientation urged by Bacon and Dee was not, when they wrote, a reality. Science conforming to that ideal did not really begin to emerge until the mid-nineteenth century. Even now science seems not to be fully prepared to endorse the unification of theory with practice that would be entailed. Pure research science is, on the one hand, characterized as the disinterested pursuit of truth (for its own sake) without regard to possible applications, and on the other urged to be necessary to ensure future technological progress. This tension in the image of science was present at the outset. And here we come to the second component.

Given that active involvement with the physical world is endorsed and incorporated into the goal of science, so that interest in successful engagement becomes an interest internal to science (a scientific interest), there is still a question of the basis from which this engagement is made and on which it is valued.

What can be said about nature, is she an enemy or a slave, an adversary or a partner in a contract . . .?[13]

For Dee and Bacon the basic relation is that between superior and inferior. Through his engagement with the world Man will demonstrate his superiority, show that he is set apart from the rest of Nature, capable by virtue of his intellect of transcending it. The standpoint at which he aims is still a standpoint outside the world, a standpoint distanced from the object of his scientific study and of his manipulation. It is this which continues to feed the conception of the objectivity of scientific knowledge as consisting in disengagement, as being disinterested, not materially conditioned, and hence as value-free. The intellect places man beyond Nature and there is a presumption that only intellectual knowledge, knowledge from a disengaged, non-human perspective can confer real power over Nature. We are thus heirs to a paradoxical conceit of knowledge characterized as neutral, disinterested, value-free, not materially conditioned, but none the less sought and valued just in so far as it confers material power. The quest for such knowledge is far from being disinterested.

But the basic relation need not be that postulated by Dee and Bacon. The exploration of alternatives is what is implicit in the many and various internal critiques of science. These accept the fundamental reorientation towards the practical and towards active involvement with the world both as a means and as an end in science, but question the basis on which the engagement is made.

Any proposal which seeks to base science on a relation between Man and Nature in which the two are put on a more equal footing will have the effect of treating Man as more fully a part of Nature. This necessarily requires abandoning the presumed availability of a non-materially conditioned viewpoint. This does not however entail abandonment of standards of objectivity. The importance of successful completion of practical projects and of reliably repeatable methods of achieving results can still be recognized even when it is realized that projects, and hence success or failure in them, are set and judged against a background of conditions which determine and limit their significance:

... there is a tendency to forget that all science is bound up with human culture in general, and that scientific findings, even those which at the moment appear the most advanced and esoteric and difficult to grasp, are meaningless outside their cultural context.[14]

Notes

1. Michel Serres, *Hermes: Literature, Science, Philosophy*, J. V. Harari and David F. Bell (eds.) (Baltimore: Johns Hopkins University Press, 1982), 99.
2. W. Newton-Smith, 'The Role of Interests in Science', in A. Phillips Griffiths (ed.), *Philosophy and Practice* (Cambridge University Press, 1984), 70.
3. M. Berman, *Social Change and Scientific Organizations: The Royal Institution 1799–1844* (London: Heinemann, 1978), xviii.
4. Isaiah Berlin, 'The Divorce between the Sciences and the Humanities' in his *Against the Current* (Oxford University Press, 1981).
5. Robert Chambers, 'Professional thinking and Rural Poverty: Putting the Last First', paper delivered to the Other Economic Summit 1985, 3.
6. Evelyn Fox Keller, 'Feminism and Science', *Signs: Journal of Women in Culture and Society*, 7/3 (1982), esp. 592 f.
7. Ilya Prigogine and Isabelle Stengers, *Order Out of Chaos: Man's New Dialogue with Nature* (London: Fontana, 1985).
8. See Newton-Smith, 'Role of Interests in Science'.
9. Charles Taylor, 'Rationality', in M. Hollis and S. Lukes (eds.), *Rationality and Relativism* (Oxford: Blackwell, 1982), 101.
10. 'Role of Interests in Science', 59.
11. For further discussion of Davy's work and its social and political context see Berman, *Social Change and Scientific Organizations* and David Albury and Joseph Schwartz, *Partial Progress* (London: Pluto Press, 1982), ch. 1.

12. These are more fully discussed in Mary Tiles, 'Mathesis and the Masculine Birth of Time', in *International Studies in the Philosophy of Science*, i (London: Routledge & Kegan Paul, 1986).
13. Serres, *Hermes: Literature, Science, Philosophy*.
14. E. Schrödinger, 'Are there Quantum Jumps?' *British Journal for the Philosophy of Science*, 3 (1952), 109.

Rethinking Standpoint Epistemology: What Is 'Strong Objectivity'?

Sandra Harding

> Feminist objectivity means quite simply situated knowledges.
>
> Donna Haraway[1]

BOTH WAYS

For almost two decades, feminists have engaged in a complex and charged conversation about objectivity. Its topics have included which kinds of knowledge projects have it, which don't, and why they don't; whether the many different feminisms need it, and if so why they do; and if it is possible to get it, how to do so. This conversation has been informed by complex and charged prefeminist writings that tend to get stuck in debates between empiricists and intentionalists, objectivists and interpretationists, and realists and social constructionists (including poststructuralists).

Most of these feminist discussions have *not* arisen from attempts to find new ways either to criticize or carry on the agendas of the disciplines. Frequently they do not take as their problematics the ones familiar with the disciplines. Instead, these conversations have emerged mainly from two different and related concerns. First, what are the causes of the immense proliferation of theoretically and empirically sound results of research in biology and the social sciences that have discovered what is not supposed to exist: rampant sexist and androcentric bias—'politics'!—in the dominant scientific (and popular) descriptions and explanations of nature and social life? To put the point another way, how should one explain the surprising fact that politically guided research projects have been able to produce less partial and distorted results of research

Reprinted with permission from Linda Alcoff and Elizabeth Potter (eds.), *Feminist Epistemologies* (New York: Routledge, 1993).

than those supposedly guided by the goal of value-neutrality? Second, how can feminists create research that is *for* women in the sense that it provides less partial and distorted answers to questions that arise from women's lives and are not only about those lives but also about the rest of nature and social relations? The two concerns are related because recommendations for future scientific practices should be informed by the best accounts of past scientific successes. That is, how one answers the second question depends on what one thinks is the best answer to the first one.

Many feminists, like thinkers in the other new social liberation movements, now hold that it is not only desirable but also possible to have that apparent contradiction in terms—socially situated knowledge. In conventional accounts, socially situated beliefs only get to count as opinions. In order to achieve the status of knowledge, beliefs are supposed to break free of—to transcend—their original ties to local, historical interests, values, and agendas. However, as Donna Haraway has put the point, it turns out to be possible 'to have *simultaneously* an account of radical historical contingency for all knowledge claims and knowing subjects, a critical practice for recognizing our own "semiotic technologies" for making meanings, and a no-nonsense commitment to faithful accounts of a "real" world. . . .'[2]

The standpoint epistemologists—and especially the feminists who have most fully articulated this kind of theory of knowledge—have claimed to provide a fundamental map or 'logic' for how to do this: 'start thought from marginalized lives' and 'take everyday life as problematic'.[3] However, these maps are easy to misread if one doesn't understand the principles used to construct them. Critics of standpoint writings have tended to refuse the invitation to 'have it both ways' by accepting the idea of real knowledge that is socially situated. Instead they have assimilated standpoint claims either to objectivism or some kind of conventional foundationalism or to ethnocentrism, relativism, or phenomenological approaches in philosophy and the social sciences.

Here I shall try to make clear how it really is a misreading to assimilate standpoint epistemologies to those older ones and that such misreadings distort or make invisible the distinctive resources that they offer. I shall do so by contrasting the grounds for knowledge and the kinds of subjects/agents of knowledge recommended by standpoint theories with those favoured by the older epistemologies. Then I shall show why it is reasonable to think that the socially situated grounds and subjects of standpoint epistemologies require and generate stronger standards for objectivity than do those that turn away from providing systematic methods for locating knowledge in history. The problem with the conventional conception of objectivity is not that it is too rigorous or too 'objectifying', as some have argued, but that it is *not rigorous or objectifying enough;* it is too weak to accomplish even the goals for which it has been designed, let alone the more difficult projects called for by feminisms and other new social movements.

FEMINIST STANDPOINT VERSUS SPONTANEOUS FEMINIST EMPIRICIST EPISTEMOLOGIES

Not all feminists who try to explain the past and learn lessons for the future of feminist research in biology and the social sciences are standpoint theorists. The distinctiveness of feminist standpoint approaches can be emphasized by contrasting them with what I shall call 'spontaneous feminist empiricist epistemology'.

By now, two forms of feminist empiricism have been articulated: the original 'spontaneous' feminist empiricism and a recent philosophical version. Originally, feminist empiricism arose as the 'spontaneous consciousness' of feminist researchers in biology and the social sciences who were trying to explain what was and what wasn't different about their research process in comparison with the standard procedures in their field. They thought that they were just doing more carefully and rigorously what any good scientist should do; the problem they saw was one of 'bad science'. Hence they did not give a special name to their philosophy of science; I gave it the name 'feminist empiricism' in *The Science Question in Feminism* to contrast feminist standpoint theory with the insistence of empiricism's proponents that sexism and androcentrism could be eliminated from the results of research if scientists would just follow more rigorously and carefully the existing methods and norms of research—which, for practising scientists, are fundamentally empiricist ones.

Recently, philosophers Helen Longino and Lynn Hankinson Nelson have developed sophisticated and valuable feminist empiricist philosophies of science (Longino calls hers 'contextual empiricism') that differ in significant respects from what most pre-feminist empiricists and probably most spontaneous feminist empiricists would think of as empiricism.[4] This is no accident, because Longino and Nelson both intend to revise empiricism, as feminists in other fields have fruitfully revised other theoretical approaches—indeed, as feminist standpoint theorists revise the theory from which they begin. Longino and Nelson incorporate into their epistemologies elements that also appear in the standpoint accounts (many would say that they have been most forcefully articulated in such accounts)—such as the inescapable but also sometimes positive influence of social values and interests in the content of science—that would be anathema to even the spontaneous feminist empiricists of the late 1970s and early 1980s as well as to their many successors today. These philosophical feminist empiricisms are constructed in opposition partly to feminist standpoint theories, partly to radical feminist arguments that exalt the feminine and essentialize 'woman's experience' (which they have sometimes attributed to standpoint theorists), and partly to the prefeminist empiricists. . . .

Here I want to show how strongly feminist reflections on scientific knowledge challenge the dominant pre-feminist epistemology and philosophy of science that are held by all of those people inside and outside science who are still wondering just what are the insights about science and knowledge that feminists have to offer. In my view, this challenge is made most strongly by feminist standpoint epistemology.

One can understand spontaneous feminist empiricism and feminist standpoint theory to be making competing arguments on two topics—scientific method and history—in order to explain in their different ways the causes of sexist and androcentric results of scientific research. As already indicated, spontaneous feminist empiricists think that insufficient care and rigour in following existing methods and norms is the cause of sexist and androcentric results of research, and it is in these terms that they try to produce plausible accounts of the successes of empirically and theoretically more adequate results of research. Standpoint theorists think that this is only part of the problem. They point out that retroactively, and with the help of the insights of the women's movement, one can see these sexist or androcentric practices in the disciplines. However, the methods and norms in the disciplines are too weak to permit researchers *systematically* to identify and eliminate from the results of research those social values, interests, and agendas that are shared by the entire scientific community or virtually all of it. Objectivity has not been 'operationalized' in such a way that scientific method can detect sexist and androcentric assumptions that are 'the dominant beliefs of an age'—that is, that are collectively (versus only individually) held. As far as scientific method goes (and feminist empiricist defences of it), it is entirely serendipitous when cultural beliefs that are assumed by most members of a scientific community are challenged by a piece of scientific research. Standpoint theory tries to address this problem by producing stronger standards for 'good method', ones that can guide more competent efforts to maximize objectivity.

With respect to history, spontaneous feminist empiricists argue that movements of social liberation such as the women's movement function much like the little boy who is the hero of the folk tale about the Emperor and his clothes. Such movements 'make it possible for people to see the world in an enlarged perspective because they remove the covers and blinders that obscure knowledge and observation'.[5] Feminist standpoint theorists agree with this assessment, but argue that researchers can do more than just wait around until social movements happen and then wait around some more until their effects happen to reach inside the processes of producing maximally objective, causal accounts of nature and social relations. Knowledge projects can find active ways incorporated into their principles of 'good method' to use history as a resource by socially situating knowledge projects in the scientifically and epistemologically most favourable historical locations. History can become the systematic provider of scientific and epistemological resources rather than an obstacle to or the 'accidental' benefactor of projects to generate knowledge.

It is spontaneous feminist empiricism's great strength that it explains the production of sexist and nonsexist results of research with only a minimal challenge to the fundamental logic of research as this is understood in scientific fields and to the logic of explanation as this is understood in the dominant philosophies of science. Spontaneous feminist empiricists try to fit feminist projects into prevailing standards of 'good science' and 'good philosophy'. . . . However, this conservatism is also this philosophy's weakness; this theory of knowledge refuses fully to address the limitations of the dominant conceptions of method and explanation and the ways the conceptions constrain and distort results of research and thought about this research even when these dominant conceptions are most rigorously respected. Nevertheless, its radical nature should not be underestimated. It argues persuasively that the sciences have been blind to their own sexist and androcentric research practices and results.

. . .

Even though standpoint arguments are most fully articulated as such in feminist writings, they appear in the scientific projects of all of the new social movements.[6]

. . .

The starting point of standpoint theory—and its claim that is most often misread—is that in societies stratified by race, ethnicity, class, gender, sexuality, or some other such politics shaping the very structure of a society, the *activities* of those at the top both organize and set limits on what persons who perform such activities can understand about themselves and the world around them. 'There are some perspectives on society from which, however well-intentioned one may be, the real relations of humans with each other and with the natural world are not visible.'[7] In contrast, the activities of those at the bottom of such social hierarchies can provide starting points for thought—for *everyone's* research and scholarship—from which humans' relations with each other and the natural world can become visible. This is because the experience and lives of marginalized peoples, as they understand them, provide particularly significant *problems to be explained* or research agendas. These experiences and lives have been devalued or ignored as a source of objectivity-maximizing questions—he answers to which are not necessarily to be found in those experiences or lives but elsewhere in the beliefs and activities of people at the centre who make policies and engage in social practices that shape marginal lives. So one's social situation enables and sets limits on what one can know; some social situations—critically unexamined dominant ones—are more limiting than others in this respect, and what makes these situations more limiting is their inability to generate the most critical questions about received belief.

It is this sense in which Dorothy Smith argues that women's experience is the 'grounds' of feminist knowledge and that such knowledge should change the discipline of sociology.[8] Women's lives (our many different lives and different experiences!) can provide the starting point for asking new, critical questions about not only those women's lives but also about men's lives and, most importantly, the causal relations between them. For example, she points out that if we start thinking from women's lives, we (anyone) can see that women are assigned the work that men do not want to do for themselves, especially the care of everyone's bodies—the bodies of men, babies, children, old people, the sick, and their own bodies. And they are assigned responsibility for the local places where those bodies exist as they clean and care for their own and others' houses and work-places.[9] This kind of 'women's work' frees men in the ruling groups to immerse themselves in the world of abstract concepts. The more successful women are at this concrete work, the more invisible it becomes to men as distinctively social labour. Caring for bodies and the places bodies exist disappears into 'nature', as, for example, in sociobiological claims about the naturalness of 'altruistic' behaviour for females and its unnaturalness for males or in the systematic reticence of many pre-feminist Marxists actually to analyse who does what in everyday sexual, emotional, and domestic work, and to integrate such analyses into their accounts of 'working class labour'. Smith argues that we should not be surprised that men have trouble seeing women's activities as part of distinctively human culture and history once we notice how invisible the social character of this work is from the perspective of their activities. She points out that if we start from women's lives, we can generate questions about why it is that it is primarily women who are assigned such activities and what the consequences are for the economy, the state, the family, the educational system, and other social institutions of assigning body and emotional work to one group and 'head' work to another.[24] These questions lead to less partial and distorted understandings of women's worlds, men's worlds, and the causal relations between them than do the questions originating only in that part of human activity that men in the dominant groups reserve for themselves—the abstract mental work of managing and administrating.

Standpoint epistemology sets the relationship between knowledge and politics at the centre of its account in the sense that it tries to provide causal accounts—to explain—the effects that different kinds of politics have on the production of knowledge. Of course, empiricism also is concerned with the effects politics has on the production of knowledge, but pre-feminist empiricism conceptualizes politics as entirely bad. Empiricism tries to purify science of all such bad politics by adherence to what it takes to be rigorous methods for the testing of hypotheses. From the perspective of standpoint epistemology, this is *far too weak a strategy* to maximize the objectivity of the results of research that empiricists desire. Thought that begins from the lives of the oppressed has no chance to get its critical questions voiced or heard within such an empiricist conception of the way to produce knowledge. Pre-feminist empiricists can only perceive such questions as the intrusion of politics into science, which therefore deteriorates the objectivity of the results of research. Spontaneous feminist empiricism, for all its considerable virtues, nevertheless contains distorting traces of these assumptions, and they block the ability of this theory of science to develop maximally strong criteria for systematic ways to maximize objectivity.

Thus the standpoint claims that all knowledge attempts are socially situated and that some of these objective social locations are better than others as starting points for knowledge projects challenge some of the most fundamental assumptions of the scientific world view and the Western thought that takes science as its model of how to produce knowledge. It sets out a rigorous 'logic of discovery' intended to maximize the objectivity of the results of research and thereby to produce knowledge that can be *for* marginalized people (and those who would know what the marginalized can know) rather than *for* the use only of dominant groups in their projects of administering and managing the lives of marginalized people.

. . .

WHAT ARE THE GROUNDS FOR KNOWLEDGE CLAIMS?[11]

The claim that asking questions, selecting problematics, by starting off thought from women's lives can provide more reliable 'grounds' for knowledge is easily misunderstood from the perspective of conventional epistemologies and philosophies of science. First of all, it denies that any questions asked by humans could be universal in the sense of expressing no particular historical values and interests. Conventional epistemological and scientific questions have excluded values and interests arising from women's lives; they are thus as socially determinate as the feminist accounts that begin from women's lives. However, second, standpoint theorists are not arguing that only 'their own' lives are to be preferred as starting points for knowledge-seeking projects. Their concern is with the positive scientific and epistemic value of marginality whether or not 'their own' lives are such marginalized ones. Thus they are not arguing that only the oppressed can produce knowledge. Men have made important contributions to feminist research and scholarship, whites to research and scholarship starting from the lives of peoples of colour, etc., as even the most cursory look at writings in these fields reveals.

Third, standpoint epistemologies provide a theory of knowledge that stands in opposition not only to the conventional universalist/absolutist ones, but also to the relativist, perspectivalist, and pluralist ones presumed to be their only alternative. Of course standpoint theories are identifying the historical or sociological relativism of all knowledge claims: different social activities lead to different interactions with nature and social relations and, thus, to different representations of the world in which culturally determinate actions intervene. However, they hold that not all such historically determinate interactions are equally revealing of each of nature's regularities and their underlying causal tendencies. They refuse epistemological or cognitive relativism. They argue that everyone can learn much about nature and social relations by posing research issues from the perspective of women's lives—something that women, too, have to learn to do. These questions arise from women's lives, but their answers are to be found elsewhere.

NEW SUBJECTS OF KNOWLEDGE

For conventional empiricist epistemology, the subject/agent of knowledge is to be invisible, disembodied; in contrast to mere opinion, real knowledge is to have no determinate historical location at all. This subject is therefore different in kind from the objects of study which are, of course, visible, and often located in determinate space/time co-ordinates. On the other hand, knowledge is initially produced by identifiable individuals or groups of them, not by cultures, genders, races, or classes. Finally, because knowledge is to be consistent and coherent, the subject of knowledge must be homogenous, unitary, and coherent; it cannot be heterogeneous, multiple and conflicted.

The subjects of knowledge for standpoint theories contrast in all four respects. First, they are embodied and visible not only in that they start from some particular set of lives, but also because conceptual frameworks are always those of a particular historical moment. Thus, these standpoint subjects of knowledge are not fundamentally different from their objects of study. This includes the objects of natural science study which are, after all, never 'bare nature' but, rather, nature-as-an-object-of-knowledge, always appearing to scientists only within specific scientific traditions and local cultures—the earth as part of nature's clockworks, or as a spaceship, for example. Third, in an important sense it is historical moments—communities and not individuals—that fundamentally produce knowledge. Finally, these subjects are multiple, heterogeneous, and contradictory. The most fruitful feminist problematics have emerged out of the gaps between the values and interests of women's lives and those that have organized the dominant conceptual frameworks; the gap is crucial. And the feminist thought 'of an age' has started from the gaps between many different kinds of women's lives and the dominant frameworks; race, class, sexuality, and other cultural conditions of women insure the fruitful heterogeneity of feminist discourses.

. . .

We could say that since standpoint analyses explain how and why the subject of knowledge always appears in scientific accounts of nature and social life as part of the object of knowledge of those accounts, standpoint approaches have had to learn to use the social situatedness of subjects of knowledge systematically as a resource for maximizing objectivity.

STANDARDS FOR MAXIMIZING OBJECTIVITY

. . .

Strong objectivity requires that the subject of knowledge be placed on the same critical, causal plane as the objects of knowledge. Thus, strong objectivity requires what we can think of as 'strong reflexivity'. This is because culturewide (or nearly culturewide) beliefs function as evidence at every stage in scientific inquiry: in the selection of problems, the formation of hypotheses, the design of research (including the organization of research communities), the collection of data, the interpretation and sorting of data, decisions about when to stop research, the way results of research are reported, and so on. The subject of knowledge—the individual and the historically located social community whose unexamined beliefs its members are likely to hold 'unknowingly', so to speak— must be considered as part of the object of knowledge from the perspective of scientific method. All of the kinds of objectivity-maximizing procedures focused on the nature and/or social relations that are the direct object of observation and reflection must also be focused on the observers and reflectors—scientists and the larger society whose assumptions they share. But a maximally critical study of scientists and their communities can be done only from the perspective of those whose lives have been marginalized by such communities. Thus, strong objectivity requires that scientists and their communities be integrated into democracy-advancing projects for scientific and epistemological reasons as well as moral and political ones.

From the perspective of such standpoint arguments, empiricism's standards appear weak; empiricism advances only the 'objectivism' that has been so widely criticized from many quarters. Objectivism impoverishes its attempts at maximizing objectivity when it turns away from the task of critically identifying all of those broad, historical social desires, interests, and values that have shaped the agendas, contents, and results of the sciences much as they shape the rest of human affairs.

Consider, first, how objectivism too narrowly operationalizes the notion of maximizing objectivity. The conception of value-free, impartial, dispassionate research is supposed to direct the identification of all social values and their elimination from the results of research, yet it has been operationalized to identify and eliminate only those social values and interests that differ among the researchers and critics who are regarded by the scientific community as competent to make such judgements. If the community of 'qualified' researchers and critics systematically excludes, for example, all African-Americans and women of all races and if the larger culture is stratified by race and gender and lacks powerful critiques of this stratification, it is not plausible to imagine that racist and sexist interests and values would be identified within a community of scientists composed entirely of people who benefit— intentionally or not—from institutionalized racism and sexism. This kind of blindness is advanced by the conventional belief that the truly scientific part of knowledge seeking—the part controlled by methods of research—occurs only in the context of justification. The context of discovery, in which problems are identified as appropriate for scientific investigation, hypotheses are formulated, key concepts are defined—this part of the scientific process is thought to be unexaminable within science by rational methods. Thus 'real science' is restricted to those processes controllable by methodological rules. The methods of science—or rather, of the special sciences—are restricted to procedures for the testing of already formulated hypotheses. Untouched by these methods are those values and interests entrenched in the very statement of what problem is to be researched and in the concepts favoured in the hypotheses that are to be tested. Recent histories of science are full of cases in which broad social assumptions stood little chance of identification or elimination through the very best research procedures of the day. Thus objectivism operationalizes the notion of objectivity in much too narrow a way to permit the achievement of the value-free research that is supposed to be its outcome.

But objectivism also conceptualizes the desired value-neutrality of objectivity too broadly. Objectivists claim that objectivity requires the elimination of *all* social values and interests from the research process and the results of research. It is clear, however, that not all social values and interests have the same bad effects upon the results of research. Democracy-advancing values have systematically generated less partial and distorted beliefs than others.

Objectivism's rather weak standards for maximizing objectivity make objectivity a mystifying notion, and its mystificatory character is largely responsible for its usefulness and its widespread appeal to dominant groups. It offers hope that scientists and science institutions, themselves admittedly historically located, can produce claims that will be regarded as objectively valid without having to examine critically their own historical commitments from which—intentionally or not—they actively construct their scientific research. It permits scientists and science institutions to be unconcerned with the origins or consequences of their problematics and practices or with the social values and interests that these problematics and practices support. It offers the false hope of enacting what Francis Bacon erroneously promised for the method of modern science: 'The course I propose for the discovery of sciences is such as leaves but little to the acuteness and strength of wits, but places all wits and understandings nearly on a level.' His 'way of discovering science goes far to level men's wits, and leaves but little to individual excellence, because it performs everything by surest rules and demonstrations'.[13] In contrast, standpoint approaches requires the strong objectivity that can take the subject as well as the object of knowledge to be a necessary object of critical, causal—scientific!—social explanations. This programme of strong reflexivity is a resource for objectivity, in contrast to the obstacle that *de facto* reflexivity has posed to weak objectivity.

Some feminists and thinkers from other liberatory knowledge projects have thought that the very notion of objectivity should be abandoned. They say that it is hopelessly tainted by its use in racist, imperialist, bourgeois, homophobic, and androcentric scientific projects. Moreover, it is tied to a theory of representation and concept of the self or subject that insists on a rigid barrier between sub-

16 Situated Knowledges: The Science Question in Feminism and the Privilege of Partial Perspective

Donna Haraway

Academic and activist feminist inquiry has repeatedly tried to come to terms with the question of what *we* might mean by the curious and inescapable term 'objectivity'. We have used a lot of toxic ink and trees processed into paper decrying what *they* have meant and how it hurts *us*. The imagined 'they' constitute a kind of invisible conspiracy of masculinist scientists and philosophers replete with grants and laboratories. The imagined 'we' are the embodied others, who are not allowed *not* to have a body, a finite point of view, and so an inevitably disqualifying and polluting bias in any discussion of consequence outside our own little circles, where a 'mass'-subscription journal might reach a few thousand readers composed mostly of science haters. At least, I confess to these paranoid fantasies and academic resentments lurking underneath some convoluted reflections in print under my name in the feminist literature in the history and philosophy of science. We, the feminists in the debates about science and technology, are the Reagan era's 'special-interest groups' in the rarified realm of epistemology, where traditionally what can count as knowledge is policed by philosophers codifying cognitive canon law. Of course, a special-interest group is, by Reaganoid definition, any collective historical subject that dares to resist the stripped-down atomism of Star Wars, hypermarket, postmodern, media-simulated citizenship.

. . .

It has seemed to me that feminists have both selectively and flexibly used and been trapped by two poles of a tempting dichotomy on the question of objectivity. Certainly I speak for myself here, and I offer the speculation that there is a collective discourse on these matters. Recent social studies of science and technology, for example, have made available a very strong social constructionist

Reprinted with permission from *Feminist Studies,* 14/3 (1988), 575–99.

argument for *all* forms of knowledge claims, most certainly and especially scientific ones.[1] According to these tempting view, no insider's perspective is privileged, because all drawings of inside-outside boundaries in knowledge are theorized as power moves, not moves toward truth. So, from the strong social constructionist perspective, why should we be cowed by scientists' descriptions of their activity and accomplishments; they and their patrons have stakes in throwing sand in our eyes. They tell parables about objectivity and scientific method to students in the first years of their initiation, but no practitioner of the high scientific arts would be caught dead *acting on* the textbook versions. Social constructionists make clear that official ideologies about objectivity and scientific method are particularly bad guides to how scientific knowledge is actually *made.* Just as for the rest of us, what scientists believe or say they do and what they really do have a very loose fit.

. . .

From this point of view, science—the real game in town—is rhetoric, a series of efforts to persuade relevant social actors that one's manufactured knowledge is a route to a desired form of very objective power. Such persuasions must take account of the structure of facts and artefacts, as well as of language-mediated actors in the knowledge game. Here, artefacts and facts are parts of the powerful art of rhetoric. Practice is persuasion, and the focus is very much on practice. All knowledge is a condensed node in an agonistic power field. The strong programme in the sociology of knowledge joins with the lovely and nasty tools of semiology and deconstruction to insist on the rhetorical nature of truth, including scientific truth. History is a story Western culture buffs tell each other; science is a contestable text and a power field; the content is the form. Period.

So much for those of us who would still like to talk about *reality* with more confidence than we allow to the Christian Right when they discuss the Second Coming and their being raptured out of the final destruction of the world. We would like to think our appeals to real worlds are more than a desperate lurch away from cynicism and an act of faith like any other cult's, no matter how much space we generously give to all the rich and always historically specific mediations through which we and everybody else must know the world. But the further I get in describing the radical social constructionist programme and a particular version of postmodernism, coupled with the acid tools of critical discourse in the human sciences, the more nervous I get. The imagery of force fields,

of moves in a fully textualized and coded world, which is the working metaphor in many arguments about socially negotiated reality for the postmodern subject, is, just for starters, an imagery of high-tech military fields, of automated academic battlefields, where blips of light called players disintegrate (what a metaphor!) each other in order to stay in the knowledge and power game. Technoscience and science fiction collapse into the sun of their radiant (ir)reality—war.[2] It shouldn't take decades of feminist theory to sense the enemy here. Nancy Hartsock got all this crystal clear in her concept of abstract masculinity.[3]
. . .

Some of us tried to stay sane in these disassembled and dissembling times by holding out for a feminist version of objectivity. Here, motivated by many of the same political desires, is the other seductive end of the objectivity problem. Humanistic Marxism was polluted at the source by its structuring theory about the domination of nature in the self-construction of man and by its closely related impotence in relation to historicizing anything women did that didn't qualify for a wage. But Marxism was still a promising resource as a kind of epistemological feminist mental hygiene that sought our own doctrines of objective vision. Marxist starting points offered a way to get to our own versions of standpoint theories, insistent embodiment, a rich tradition of critiquing hegemony without disempowering positivisms and relativisms and a way to get to nuanced theories of mediation. Some versions of psychoanalysis were of aid in this approach, especially anglophone object relations theory, which maybe did more for US socialist feminism for a time than anything from the pen of Marx or Engels, much less Althusser or any of the late pretenders to sonship treating the subject of ideology and science.

Another approach, 'feminist empiricism', also converges with feminist uses of Marxian resources to get a theory of science which continues to insist on legitimate meanings of objectivity and which remains leery of a radical constructivism conjugated with semiology and narratology.[4] Feminists have to insist on a better account of the world; it is not enough to show radical historical contingency and modes of construction for everything. Here, we, as feminists, find ourselves perversely conjoined with the discourse of many practising scientists, who, when all is said and done, mostly believe they are describing and discovering things *by means of* all their constructing and arguing. Evelyn Fox Keller has been particularly insistent on this fundamental matter, and Sandra Harding calls the goal

251

of these approaches a 'successor science'. Feminists have stakes in a successor science project that offers a more adequate, richer, better account of a world, in order to live in it well and in critical, reflexive relation to our own as well as others' practices of domination and the unequal parts of privilege and oppression that make up all positions. In traditional philosophical categories, the issue is ethics and politics perhaps more than epistemology.

So, I think my problem, and 'our' problem, is how to have *simultaneously* an account of radical historical contingency for all knowledge claims and knowing subjects, a critical practice for recognizing our own 'semiotic technologies' for making meanings, *and* a no-nonsense commitment to faithful accounts of a 'real' world, one that can be partially shared and that is friendly to earthwide projects of finite freedom, adequate material abundance, modest meaning in suffering, and limited happiness.

. . .

Natural, social, and human sciences have always been implicated in hopes like these. Science has been about a search for translation, convertibility, mobility of meanings, and universality . . . There is, finally, only one equation. That is the deadly fantasy that feminists and others have identified in some versions of objectivity, those in the service of hierarchical and positivist orderings of what can count as knowledge. That is one of the reasons the debates about objectivity matter, metaphorically and otherwise. Immortality and omnipotence are not our goals. But we could use some enforceable, reliable accounts of things not reducible to power moves and agonistic, high-status games of rhetoric or to scientistic, positivist arrogance. This point applies whether we are talking about genes, social classes, elementary particles, genders, races, or texts; the point applies to the exact, natural, social, and human sciences, despite the slippery ambiguities of the words 'objectivity' and 'science' as we slide around the discursive terrain. In our efforts to climb the greased pole leading to a usable doctrine of objectivity, I and most other feminists in the objectivity debates have alternatively, or even simultaneously, held on to both ends of the dichotomy, a dichotomy which Harding describes in terms of successor science projects versus postmodernist accounts of difference and which I have sketched in this essay as radical constructivism versus feminist critical empiricism. It is, of course, hard to climb when you are holding on to both ends of a pole, simultaneously or alternatively. It is, therefore, time to switch metaphors.

THE PERSISTENCE OF VISION

I would like to proceed by placing metaphorical reliance on a much maligned sensory system in feminist discourse: vision. Vision can be good for avoiding binary oppositions. I would like to insist on the embodied nature of all vision and so reclaim the sensory system that has been used to signify a leap out of the marked body and into a conquering gaze from nowhere. This is the gaze that mythically inscribes all the marked bodies, that makes the unmarked category claim the power to see and not be seen, to represent while escaping representation. This gaze signifies the unmarked positions of Man and White, one of the many nasty tones of the world 'objectivity' to feminist ears in scientific and technological, late-industrial, militarized, racist, and male-dominant societies, that is, here, in the belly of the monster, in the United States in the late 1980s. I would like a doctrine of embodied objectivity that accommodates paradoxical and critical feminist science projects: Feminist objectivity means quite simply *situated knowledges*.

The eyes have been used to signify a perverse capacity—honed to perfection in the history of science tied to militarism, capitalism, colonialism, and male supremacy—to distance the knowing subject from everybody and everything in the interests of unfettered power. The instruments of visualization in multinationalist, postmodernist culture have compounded these meanings of disembodiment. The visualizing technologies are without apparent limit. The eye of any ordinary primate like us can be endlessly enhanced by sonography systems, magnetic reasonance imagining, artificial intelligence-linked graphic manipulation systems, scanning electron microscopes, computed tomography scanners, colour-enhancement techniques, satellite surveillance systems, home and office video display terminals, cameras for every purpose from filming the mucous membrane lining the gut cavity of a marine worm living in the vent gases on a fault between continental plates to mapping a planetary hemisphere elsewhere in the solar system. Vision in this technological feast becomes unregulated gluttony; all seems not just mythically about the god trick of seeing everything from nowhere, but to have put the myth into ordinary practice. And like the god trick, this eye fucks the world to make techno-monsters. Zoe Sofoulis calls this the cannibaleye of masculinist extraterrestrial projects for excremental second birthing.
. . .

I would like to suggest how our insisting metaphorically on the particularity and embodiment of all vision (although not necessarily organic embodiment and including technological mediation), and not giving in to the tempting myths of vision as a route to disembodiment and second-birthing allows us to construct a usable, but not an innocent, doctrine of objectivity. I want a feminist writing of the body that metaphorically emphasizes vision again, because we need to reclaim that sense to find our way through all the visualizing tricks and powers of modern sciences and technologies that have transformed the objectivity debates. We need to learn in our bodies, endowed with primate colour and stereoscopic vision, how to attach the objective to our theoretical and political scanners in order to name where we are and are not, in dimensions of mental and physical space we hardly know how to name. So, not so perversely, objectivity turns out to be about particular and specific embodiment and definitely not about the false vision promising transcendence of all limits and responsibility. The moral is simple: only partial perspective promises objective vision. All Western cultural narratives about objectivity are allegories of the ideologies governing the relations of what we call mind and body, distance and responsibility. Feminist objectivity is about limited location and situated knowledge, not about transcendence and splitting of subject and object. It allows us to become answerable for what we learn how to see.

These are lessons that I learned in part walking with my dogs and wondering how the world looks without a fovea and very few retinal cells for colour vision but with a huge neural processing and sensory area for smells. It is a lesson available from photographs of how the world looks to the compound eyes of an insect or even from the camera eye of a spy satellite or the digitally transmitted signals of space probe-perceived differences 'near' Jupiter that have been transformed into coffee table colour photographs. The 'eyes' made available in modern technological sciences shatter any idea of passive vision; these prosthetic devices show us that all eyes, including our own organic ones, are active perceptual systems, building on translations and specific *ways* of seeing, that is, ways of life. There is no unmediated photograph or passive camera obscura in scientific accounts of bodies and machines; there are only highly specific visual possibilities, each with a wonderfully detailed, active, partial way of organizing worlds. All these pictures of the world should not be allegories of infinite mobility and interchangeability but of elaborate specificity and difference and the loving care people might take to learn how to see faithfully from another's point of view, even when the other is our own machine. That's not alienating distance; that's a *possible* allegory for feminist versions of objectivity. Understanding how these visual systems work, technically, socially, and psychically, ought to be a way of embodying feminist objectivity.

Many currents in feminism attempt to theorize grounds for trusting especially the vantage points of the subjugated; there is good reason to believe vision is better from below the brilliant space platforms of the powerful.[5] Building on that suspicion, this essay is an argument for situated and embodied knowledges and an argument against various forms of unlocatable, and so irresponsible, knowledge claims. Irresponsible means unable to be called into account. There is a premium on establishing the capacity to see from the peripheries and the depths. But here there also lies a serious danger of romanticizing and/or appropriating the vision of the less powerful while claiming to see from their positions. To see from below is neither easily learned nor unproblematic, even if 'we' 'naturally' inhabit the great underground terrain of subjugated knowledges. The positionings of the subjugated are not exempt from critical re-examination, decoding, deconstruction, and interpretation; that is, from both semiological and hermeneutic modes of critical inquiry. The standpoints of the subjugated are not 'innocent' positions. On the contrary, they are preferred because in principle they are least likely to allow denial of the critical and interpretive core of all knowledge. They are knowledgeable of modes of denial through repression, forgetting, and disappearing acts—ways of being nowhere while claiming to see comprehensively. The subjugated have a decent chance to be on to the god trick and all its dazzling—and, therefore, blinding—illuminations. 'Subjugated' standpoints are preferred because they seem to promise more adequate, sustained, objective, transforming accounts of the world. But *how* to see from below is a problem requiring at least as much skill with bodies and language, with the mediations of vision, as the 'highest' technoscientific visualizations.

Such preferred positioning is as hostile to various forms of relativism as to the most explicitly totalizing versions of claims to scientific authority. But the alternative to relativism is not totalization and single vision, which is always finally the unmarked category whose power depends on systematic narrowing and obscuring. The alternative to relativism is partial, locatable, critical knowledges sustaining the possibility of webs of connections called solidarity in politics and shared conversations in epistemology. Relativism is a way of being nowhere while claiming to be everywhere equally. The 'equality' of positioning is a denial of responsibility and critical inquiry. Relativism is the perfect mirror twin of totalization in the ideologies of objectivity; both deny the stakes in location, embodiment, and partial perspective; both make it impossible to see well. Relativism and totalization are both 'god tricks' promising vision from everywhere and nowhere equally and fully, common myths in rhetorics surrounding Science. But it is precisely in the politics and epistemology of partial perspectives that the possibility of sustained, rational, objective inquiry rests.

. . .

A commitment to mobile positioning and to passionate detachment is dependent on the impossibility of entertaining innocent 'identity' politics and epistemologies as strategies for seeing from the standpoints of the subjugated in order to see well. One cannot 'be' either a cell or molecule—or a woman, colonized person, labourer, and so on—if one intends to see and see from these positions critically. 'Being' is much more problematic and contingent. Also, one cannot relocate in any possible vantage point without being accountable for that movement. Vision is *always* a question of the power to see—and perhaps of the violence implicit in our visualizing practices. With whose blood were my eyes crafted? These points also apply to testimony from the position of 'oneself'. We are not immediately present to ourselves. Self-knowledge requires a semiotic-material technology to link meanings and bodies. Self-identity is a bad visual system. Fusion is a bad strategy of positioning. The boys in the human sciences have called this doubt about self-presence the 'death of the subject' defined as a single ordering point of will and consciousness. That judgement seems bizarre to me. I prefer to call this doubt the opening of non-isomorphic subjects, agents, and territories of stories unimaginable from the vantage point of the cyclopean, self-satiated eye of the master subject. The Western eye has fundamentally been a wandering eye, a travelling lens. These peregrinations have often been violent and insistent on having mirrors for a conquering self—but not always. Western feminists also *inherit* some skill in learning to participate in revisualizing worlds turned upside down in earth-transforming challenges to the views of the masters. All is not to be done from scratch.

The split and contradictory self is the one who can interrogate positionings and be accountable, the one who can construct and join rational conversations and fantastic imaginings that change history. Splitting, not being, is the privileged image for feminist epistemologies of scientific knowledge. 'Splitting' in this context should be about heterogeneous multiplicities that are simultaneously salient and incapable of being squashed into isomorphic slots or cumulative lists. This geometry pertains within and among subjects. Subjectivity is multidimensional; so, therefore, is vision. The knowing self is partial in all its guises, never finished, whole, simply there and original; it is always constructed and stitched together imperfectly, and *therefore* able to join with another, to see together without claiming to be another. Here is the promise of objectivity: a scientific knower seeks the subject position, not of identity, but of objectivity, that is, partial connection. There is no way to 'be' simultaneously in all, or wholly in any, of the privileged (i.e. subjugated) positions structured by gender, race, nation, and class. And that is a short list of critical positions. The search for such a 'full' and total position is the search for the fetishized perfect subject of oppositional history, sometimes appearing in feminist theory as the essentialized Third World Woman.[6] Subjugation is not grounds for an ontology; it might be a visual clue. Vision requires instruments of vision; an optics is a politics of positioning. Instruments of vision mediate standpoints; there is no immediate vision from the standpoints of the subjugated. Identity, including self-identity, does not produce science; critical positioning does, that is, objectivity. Only those occupying the positions of the dominators are self-identical, unmarked, disembodied, unmediated, transcendent, born again. It is unfortunately possible for the subjugated to lust for and even scramble into that subject position—and then disappear from view. Knowledge from the point of view of the unmarked is truly fantastic, distorted, and irrational. The only position from which objectivity could not possibly be practised and honoured is the standpoint of the master, the Man, the One God, whose Eye produces, appropriates, and orders all difference. No one ever accused the god of monotheism of objectivity, only of indifference. The god trick is self-identical, and we have mistaken that for creativity and knowledge, omniscience even.

. . .

I am arguing for politics and epistemologies of location, positioning, and situating, where partiality and not universality is the condition of being heard to make rational knowledge claims. These are claims on people's lives. I am arguing for the view from a body, always a complex, contradictory, structuring, and structured body, versus the view from above, from nowhere, from simplicity. Only the god trick is forbidden. Here is a criterion for deciding the science question in militarism, that dream science/technology of perfect language, perfect communication, final order.

Feminism loves another science: the sciences and politics of interpretation, translation, stuttering, and the partly understood. Feminism is about the sciences of the multiple subject with (at least) double vision. Feminism is about a critical vision consequent upon a critical positioning in unhomogeneous gendered social space. Translation is always interpretive, critical, and partial. Here is a ground for conversation, rationality, and objectivity—which is power-sensitive, not pluralist, 'conversation'. It is not even the mythic cartoons of physics and mathematics—incorrectly caricatured in antiscience ideology as exact, hypersimple knowledges— that have come to represent the hostile other to feminist paradigmatic models of scientific knowledge, but the dreams of the perfectly known in high-technology, permanently militarized scientific productions and positionings, the god trick of a Star Wars paradigm of rational knowledge. So location is about vulnerability; location resists the politics of closure, finality, or to borrow from Althusser, feminist objectivity resists 'simplification in the last instance'. That is because feminist embodiment resists fixation and is insatiably curious about the webs of differential positioning. There is no single feminist standpoint because our maps require too many dimensions for that metaphor to ground our visions. But the feminist standpoint theorists' goal of an epistemology and politics of engaged, accountable positioning remains eminently potent. The goal is better accounts of the world, that is, 'science'.

Above all, rational knowledge does not pretend to disengagement: to be from everywhere and so nowhere, to be free from interpretation, from being represented, to be fully self-contained or fully formalizable. Rational knowledge is a process of ongoing critical interpretation among 'fields' of interpreters and decoders. Rational knowledge is power-sensitive conversation.[7] Decoding and transcoding plus translation and criticism; all are necessary. So science becomes the paradigmatic model, not of closure, but of that which is contestable and contested. Science becomes the myth, not of what escapes human agency and responsibility in a realm above the fray, but, rather, of accountability and responsibility for translations and solidarities linking the cacophonous visions and visionary voices that characterize the knowledges of the subjugated. A splitting of senses, a confusion of voice and sight, rather than clear and distinct ideas, becomes the metaphor for the ground of the rational. We seek not the knowledges ruled by phallogocentrism (nostalgia for the presence of the one true Word) and disembodied vision. We seek those ruled by partial sight and limited voice—not partiality for its own sake but, rather, for the sake of the connections and unexpected openings situated knowledges make possible. Situated knowledges are about communities, not about isolated individuals. The only way to find a larger vision is to be somewhere in particular. The science question in feminism is about objectivity as positioned rationality. Its images are not the products of escape and transcendence of limits (the view from above) but the joining of partial views and halting voices into a collective subject position that promises a vision of the means of ongoing finite embodiment, of living within limits and contradictions—of views from somewhere.

OBJECTS AS ACTORS: THE APPARATUS OF BODILY PRODUCTION

. . .

It seems clear that feminist accounts of objectivity and embodiment—that is, of a world—of the kind sketched in this essay require a deceptively simple manœuvre within inherited Western analytical traditions, a manœuvre begun in dialectics but stopping short of the needed revisions. Situated knowledges require that the object of knowledge be pictured as an actor and agent, not as a screen or a ground or a resource, never finally as slave to the master that closes off the dialectic in his unique agency and his authorship of 'objective' knowledge. The point is paradigmatically clear in critical approaches to the social and human sciences, where the agency of people studied itself transforms the entire project of producing social theory. Indeed, coming to terms with the agency of the 'objects' studied is the only way to avoid gross error and false knowledge of many kinds in these sciences. But the same point must apply to the other knowledge projects called sciences. A corollary of the insistence that ethics and politics covertly or overtly provide the bases for objectivity in the sciences as a heterogeneous whole, and not just in the social sciences, is granting the status of agent/actor to the 'objects' of the world. Actors come in many and wonderful forms. Accounts of a 'real' world do not, then, depend on a logic of 'discovery' but on a power-charged social relation of 'conversation'. The world neither speaks itself nor disappears in favour of a master decoder. The codes of the world are not still, waiting only to be read. The world is not raw material for humanization; the thorough attacks on humanism, another branch of 'death of the subject' discourse, have made this point quite clear. In some critical sense that is crudely hinted at by the clumsy category of the social or of agency, the world encountered in knowledge projects is an active entity. In so far as a scientific account has been able to engage this dimension of the world as object of knowledge, faithful knowledge can be imagined and can make claims on us. But no particular doctrine of representation or decoding or discovery guarantees anything. The approach I am recommending is not a version of 'realism', which has proved a rather poor way of engaging with the world's active agency.

My simple, perhaps simple-minded, manœuvre is obviously not new in Western philosophy, but it has a special feminist edge to it in relation to the science question in feminism and to the linked question of gender as situated difference and the question of female embodiment. Ecofeminists have perhaps been most insistent on some version of the world as active subject, not as resource to be mapped and appropriated in bourgeois, Marxist, or masculinist projects. Acknowledging the agency of the world in knowledge makes room for some unsettling possibilities, including a sense of the world's independent sense of humour. Such a sense of humour is not comfortable for humanists and others committed to the world as resource. There are, however, richly evocative figures to promote feminist visualizations of the world as witty agent. We need not lapse into appeals to a primal mother resisting her translation into resource. The Coyote or Trickster, as embodied in Southwest native American accounts, suggests the situation we are in when we give up mastery but keep searching for fidelity, knowing all the while that we will be hoodwinked. I think these are useful myths for scientists who might be our allies. Feminist objectivity makes room for surprises and ironies at the heart of all knowledge production; we are not in charge of the world. We just live here and try to strike up non-innocent conversations by means of our prosthetic devices, including our visualization technologies. No wonder science fiction has been such a rich writing practice in recent feminist theory. I like to see feminist theory as a reinvented coyote discourse obligated to its sources in many heterogeneous accounts of the world.

Another rich feminist practice in science in the last couple of decades illustrates particularly well the 'activation' of the previously passive categories of objects of knowledge. This activation permanently problematizes binary distinctions like sex and gender, without eliminating their strategic utility. I refer to the reconstructions in primatology (especially, but not only, in women's practice as primatologists, evolutionary biologists, and behavioural ecologists) of what may count as sex, especially as female sex, in scientific accounts.[8] The *body*, the object of biological discourse, becomes a most engaging being. Claims of biological determinism can never be the same again. When female 'sex' has been so thoroughly retheorized and revisualized that it emerges as practically indistinguishable from 'mind', something basic has happened to the categories of biology. The biological female peopling current biological behavioural accounts has almost no passive properties left. She is structuring and active in every respect; the 'body' is an agent, not a resource. Difference is theorized *biologically* as situational, not intrinsic, at every level from gene to foraging pattern, thereby fundamentally changing the biological politics of the body. The relations between sex and gender need to be categorically reworked within these frames of knowledge. I would like to suggest that this trend in explanatory strategies in biology is an allegory for interventions faithful to projects of feminist objectivity. The point is not that these new pictures of the biological female are simply true or not open to contestation and conversation quite the opposite. But these pictures foreground knowledge as situated conversation at every level of its articulation. The boundary between animal and human is one of the stakes in this allegory, as is the boundary between machine and organism.

So I will close with a final category useful to a feminist theory of situated knowledges: the apparatus of bodily production. In her analysis of the production of the poem as an object of literary value, Katie King offers tools that clarify matters in the objectivity debates among feminists. King suggests the term 'apparatus of literary production' to refer to the emergence of literature at the intersection of art, business, and technology. The apparatus of literary production is a matrix from which 'literature' is born. Focusing on the potent object of value called the 'poem', King applies her analytic framework to the relation of women and writing technologies.[9] I would like to adapt her work to understanding the generation—the actual production and reproduction—of bodies and other objects of value in scientific knowledge projects. At first glance, there is a limitation to using King's scheme inherent in the 'facticity' of biological discourse that is absent from literary discourse and its knowledge claims. Are biological bodies 'produced' or 'generated' in the same strong sense as poems? From the early stirrings of Romanticism in the late eighteenth century, many poets and biologists have believed that poetry and organisms are siblings. *Frankenstein* may be read as a meditation on this proposition. I continue to believe in this potent proposition but in a postmodern and not a Romantic manner. I wish to translate the ideological dimensions of 'facticity' and 'the organic' into a cumbersome entity called a 'material-semiotic actor'. This unwieldy term is intended to portray the object of knowledge as an active, meaning-generating part of apparatus of bodily production, without *ever* implying the immediate presence of such objects or, what is the same thing, their final or unique determination of what can count as objective knowledge at a par-ticular historical juncture. Like 'poems', which are sites of literary production where language too is an actor independent of inten-tions and authors, bodies as objects of knowledge are material-semiotic generative nodes. Their *boundaries* materialize in social interaction. Boundaries are drawn by mapping practices; 'objects' do not pre-exist as such. Objects are boundary projects. But bound-aries shift from within; boundaries are very tricky. What bound-aries provisionally contain remains generative, productive of meanings and bodies. Siting (sighting) boundaries is a risky prac-tice.

Objectivity is not about disengagement but about mutual *and* usually unequal structuring, about taking risks in a world where 'we' are permanently mortal, that is, not in 'final' control. We have, finally, no clear and distinct ideas. The various contending biologi-cal bodies emerge at the intersection of biological research and writing, medical and other business practices, and technology, such as the visualization technologies enlisted as metaphors in this essay. But also invited into that node of intersection is the analogue to the lively languages that actively intertwine in the production of liter-ary value: the coyote and the protean embodiments of the world as witty agent and actor. Perhaps the world resists being reduced to mere resource because it is—not mother/matter/mutter—but coy-ote, a figure of the always problematic, always potent tie between meaning and bodies. Feminist embodiment, feminist hopes for partiality, objectivity, and situated knowledges, turn on conversa-tions and codes at this potent node in fields of possible bodies and meanings. Here is where science, science fantasy and science fiction converge in the objectivity question in feminism. Perhaps our

hopes for accountability, for politics, for ecofeminism, turn on revisioning the world as coding trickster with whom we must learn to converse.

Notes

1. For example, see Karin Knorr-Cetina and Michael Mulkay (eds.), *Science Observed: Perspectives on the Social Study of Science* (London: Sage, 1983); Wiebe E. Bijker, Thomas P. Hughes, and Trevor Pinch (eds.), *The Social Construction of Technological Systems* (Cambridge: MIT Press, 1987); and esp. Bruno Latour's *Les microbes, guerre et paix, suivi de irréductions* (Paris: Métailié, 1984) and *The Pasteurization of France, Followed by Irreductions: A Politico-Scientific Essay* (Cambridge: Harvard University Press, 1988). . . .

2. In 'Through the Lumen: Frankenstein and the Optics of Re-Origination' (Ph.D. diss. University of California at Santa Cruz, 1988), Zoe Sofoulis has produced a dazzling (she will forgive me the metaphor) theoretical treatment of techno-science, the psychoanalysis of science fiction culture, and the metaphorics of extraterrestrialism. . . . My essay was revised in dialogue with Sofoulis's arguments and metaphors in her dissertation.

3. Nancy Harstock, *Money, Sex, and Power: An Essay on Domination and Community* (Boston: Northeastern University Press, 1984).

4. Harding, 24–26, 161–62.

5. See Hartsock, 'The Feminist Standpoint: Developing the Ground for a Specifically Feminist Historical Materialism'; and Chela Sandoral, *Yours in Struggle: Women Respond to Racism* (Oakland: Center for Third World Organizing, n.d.); Harding; and Gloria Anzaldúa, *Borderlands/La Frontera* (San Francsico: Spinsters/Aunt Lute, 1987).

6. Chandra Mohanty, 'Under Western Eyes', *Boundary*, 2 and 3 (1984): 333–58.

7. Katie King, 'Canons without Innocence' (Ph.D. diss., University of California at Santa Cruz, 1987).

8. Donna Haraway, *Primate Visions: Gender, Race, and Nature in the World of Modern Science* (New York: Routledge & Kegan Paul, 1989).

9. Katie King, prospectus for 'The Passing Dreams of Choice . . . Once Before and After: Audre Lorde and the Apparatus of Literary Production' (MS, University of Maryland, College Park, Maryland, 1987).

Subjects, Power, and Knowledge: Description and Prescription in Feminist Philosophies of Science

Helen E. Longino

PROLOGUE

Feminists, faced with traditions in philosophy and in science that are deeply hostile to women, have had practically to invest new and more appropriate ways of knowing the world. These new ways have been less invention out of whole cloth than the revival or re-evaluation of alternative or suppressed traditions. They range from the celebration of insight into nature through identification with it to specific strategies of survey research in the social sciences. Natural scientists and lay persons anxious to see the sciences change have celebrated Barbara McClintock's loving identification with various aspects of the plants she studied, whether whole organism or its chromosomal structure revealed under the microscope. Social scientists from Dorothy Smith to Karen Sacks have stressed designing research *for* rather than merely about women, a goal that requires attending to the specificities of women's lives and consulting research subjects themselves about the process of gathering information about them. Such new ways of approaching natural and social phenomena can be seen as methods of discovery, ways of getting information about the natural and social worlds not available via more traditional experimental or investigative methods.

Feminists have rightly pointed out the blinders imposed by the philosophical distinction between discovery and justification; a theory of scientific inquiry that focuses solely on the logic of justification neglects the selection processes occurring in the context of discovery that limit what we get to know about. . . .

Nevertheless, ignoring the context of justification for the context of discovery is equally problematic. I wish in this essay to explore

Reprinted with permission from Linda Alcoff and Elizabeth Potter (eds.), *Feminist Epistemologies* (New York: Routledge, 1993).

some of the tensions between descriptivism and normativism (or prescriptivism) in the theory of knowledge, arguing that although many of the most familiar feminist accounts of science have helped us to redescribe the process of knowledge (or belief) acquisition, they stop short of an adequate normative theory. However, these accounts do require a new approach in normative epistemology because of their redescription.

. . .

Feminists have also been struck by the interlocking character of several aspects of knowledge and power in the sciences. Women have been excluded from the practice of science, even as scientific inquiry gets described both as a masculine activity and as demonstrating women's unsuitability to engage in it, whether because of our allegedly deficient mathematical abilities or our insufficient independence. Some of us notice the location of women in the production of the artefacts made possible by new knowledge: swift and nimble fingers on the microelectronics assembly line. Others notice the neglect of women's distinctive health issues by the biomedical sciences, even as new techniques for preserving the fetuses they carry are introduced into hospital delivery rooms. The sciences become even more suspect as analysis of their metaphors (for example, in cell biology and in microbiology) reveals an acceptance (and hence reinforcement) of the cultural identification of the male with activity and of the female with passivity. Finally, feminists have drawn a connection between the identification of nature as female and the scientific mind as male and the persistent privileging of explanatory models constructed around relations of unidirectional control over models constructed around relations of interdependence. Reflection on this connection has prompted feminist critics to question the very idea of a scientific method capable of adjudicating the truth or probability of theories in a value-neutral way.

Although the sciences have increased human power over natural processes, they have, according to this analysis, done so in a lopsided way, systematically perpetuating women's cognitive and political disempowerment (as well as that of other groups marginalized in relation to the Euro-American drama). One obvious question, then, is whether this appropriation of power is an intrinsic feature of science or whether it is an incidental feature of the sciences as practised in the modern period, a feature deriving from the social structures within which the sciences have developed. A second question is whether it is possible to seek and possess empowering knowledge without expropriating the power of others. Is seeking

knowledge inevitably an attempt at domination? And are there criteria of knowledge other than the ability to control the phenomena about which one seeks knowledge? Feminists have answered these questions in a number of ways. I will review some of these before outlining my own answer.

FEMINIST EPISTEMOLOGICAL STRATEGIES 1: CHANGING THE SUBJECT

Most traditional philosophy of science (with the problematic exception of Descartes's) has adopted some form of empiricism. Empiricism's silent partner has been a theory of the subject, that is, of the knower.[1] The paradigmatic knower in Western epistemology is an individual—an individual who, in several classic instances, has struggled to free himself from the distortions in understanding and perception that result from attachment. Plato, for example, maintained that knowledge of the good is possible only for those whose reason is capable of controlling their appetites and passions, some of which have their source in bodily needs and pleasures and others of which have their source in our relations with others. The struggle for epistemic autonomy is even starker for Descartes, who suspends belief in all but his own existence in order to recreate a body of knowledge cleansed of faults, impurities, and uncertainties. For Descartes, only those grounds available to a single, unattached, disembodied mind are acceptable principles for the construction of a system of beliefs. Most subsequent epistemology has granted Descartes's conditions and disputed what those grounds are and whether any proposed grounds are sufficient grounds for knowledge. Descartes's creation of the radically and in principle isolated individual as the ideal epistemic agent has for the most part gone unremarked.[2] Locke, for example, adopts the Cartesian identification of the thinking subject with the disembodied soul without even remarking upon the individualism of the conception he inherits and then struggles with the problem of personal identity. Explicitly or implicitly in modern epistemology, whether rationalist or empiricist, the individual consciousness that is the subject of knowledge is transparent to itself, operates according to principles that are independent of embodied experience, and generates knowledge in a value-neutral way.

One set of feminist epistemological strategies, sometimes

described as modifications or rejections of empiricism, can also, and perhaps better, be described as changing the subject. I will review three such strategies of replacement, arguing that although they enrich our understanding of how we come to have the beliefs we have and so are more descriptively adequate than the theories they challenge, they fall short of normative adequacy. The strategies identify the problems of contemporary science as resulting from male or masculinist bias. Each strategy understands both the bias and its remedy differently. One holds out the original ideal of uncontaminated or unconditioned subjectivity. A second identifies bias as a function of social location. A third identifies bias in the emotive substructure produced by the psychodynamics of individuation.

Feminist empiricism has by now taken a number of forms. That form discussed and criticized by Sandra Harding is most concerned with those fields of scientific research that have misdescribed or misanalysed women's lives and bodies. . . .

From this perspective, certain areas of science having to do with sex and gender are deformed by gender ideology, but the methods of science are not themselves masculinist and can be used to correct the errors produced by ideology. The ideal knower is still the purified mind, and epistemic or cognitive authority inheres in this purity. This strategy, as Harding has observed, is not effective against those research programmes that feminists find troublesome but that cannot be faulted by reference to the standard methodological precepts of scientific inquiry. I have argued, for example, that a critique of research on the influence of prenatal gonadal hormones on behavioural sex differences that is limited to methodological critique of the data fails to bring out the role of the explanatory model that both generates the research and gives evidential relevance to that data.[3]

Another approach is, therefore, the standpoint approach. There is no one position from which value-free knowledge can be developed, but some positions are better than others.
. . .

By valorizing the perspectives uniquely available to those who are socially disadvantaged, standpoint theorists turn the table on traditional epistemology; the ideal epistemic agent is not an unconditioned subject but the subject conditioned by the social experiences of oppression. The powerless are those with epistemic legitimacy, even if they lack the power that could turn that legitimacy into authority. One of the difficulties of the standpoint approach comes into high relief, however, when it is a women's or a feminist standpoint that is in question. Women occupy many social locations in a racially and economically stratified society. If genuine or better knowledge depends on the correct or a more correct standpoint, social theory is needed to ascertain which of these locations is the epistemologically privileged one. But in a standpoint epistemology, a standpoint is needed to justify such a theory. What is that standpoint and how do we identify *it*? If no single standpoint is privileged, then either the standpoint theorist must embrace multiple and incompatible knowledge positions or offer some means of transforming or integrating multiple perspectives into one. Both of these moves require either the abandonment or the supplementation of standpoint as an epistemic criterion.

Standpoint theory faces another problem as well. It is by now commonplace to note that standpoint theory was developed by and for social scientists. It has been difficult to see what its implications for the natural sciences might be. But another strategy has seemed more promising. Most standpoint theorists locate the epistemic advantage in the productive/reproductive experience of the oppressed whose perspective they champion. A different change of subject is proposed by those identifying the problems with science as a function of the psychodynamics of individuation. Evelyn Fox Keller has been asking, among other things, why the scientific community privileges one kind of explanation or theory over others. In particular she has asked why, when both linear reductionist and interactionist perspectives are available, the scientific community has preferred the linear or 'master molecule' theory that understands a natural process as controlled by a single dominant factor. This question was made vivid by her discussion of her own research on slime mould aggregation and the fate of Barbara McClintock's work on genetic transposition.[4]

Keller's original response, spelled out in *Reflections on Gender and Science*, involved an analysis of the traditional ideal of scientific objectivity, which she understood as the ideal of the scientist's detachment from the object of study.[5] . . .

She, therefore, proposed an alternative conceptualization of autonomy, contrasting static autonomy with what she called dynamic autonomy, an ability to move in and out of intimate connection with the world. Dynamic autonomy provides the emotional substructure for an alternative conception of objectivity: dynamic objectivity. The knower characterized by dynamic objectivity, in contrast to the knower characterized by static objectivity, does not seek power over phenomena but acknowledges instead the ways in which knower and phenomena are in relationship as well as the ways in which phenomena themselves are complexly interdependent. . . .

Both standpoint theory and the psychodynamic perspective suggest the inadequacy of an ideal of a pure transparent subjectivity that registers the world as it is in itself (or, for Kantians, as structured by universal conditions of apperception or categories of understanding). I find it most useful to read them as articulating special instances of more general descriptive claims that subjectivity is conditioned by social and historical location and that our cognitive efforts have an ineluctably affective dimension. Classical standpoint theory identifies relation to production/reproduction as the key, but there are multiple, potentially oppositional relations to production/reproduction in a complex society, and there are other kinds of social relation and location that condition subjectivity. For example, one of the structural features of a male-dominant society is asymmetry of sexual access. Men occupy a position of entitlement to women's bodies, whereas women, correspondingly, occupy the position of that to which men are entitled. Complications of the asymmetry arise in class- and race-stratified societies. There may be other structural features as well, such as those related to the institutions of heterosexuality, that condition subjectivity. Because each individual occupies a location in a multidimensional grid marked by numerous interacting structures of power asymmetry, the analytical task is not to determine which is epistemically most adequate. Rather, the task is to understand how these complexly conditioned subjectivities are expressed in action and belief. I would expect that comparable complexity can be introduced into the psychodynamic account.

Treating subjectivity as variably conditioned and cognition as affectively modulated opens both opportunities and problems. The opportunities are the possibilities of understanding phenomena in new ways; by recognizing that mainstream accounts of natural processes have been developed from particular locations and reflect particular affective orientations, we can entertain the possibility that quite different accounts might emerge from other locations with the benefit of different emotional orientations. Although either transferring or diffusing power, the strategies discussed so far have in common a focus on the individual epistemic agent, on the autonomous subject. (The subject in the second and third approaches comes to be in a social context and as a consequence of

269

social interactions, but its knowledge is still a matter of some relation between it and the subject matter.) The standpoint and psychodynamically based theories recommend certain new positions and orientations as superior to others but fail to explain how we are to decide or to justify decisions between what seem to be conflicting claims about the character of some set of natural processes. On what grounds can one social location or affective orientation be judged epistemically superior to another?

. . .

Feminist science critics have provided analyses of the context of discovery that enable us to see how social values, including gender ideology in various guises, could be introduced into science. Some theories that have done so go on to recommend an alternate subject position as epistemically superior. But arguments are missing—and it's not clear that any particular subject position could be adequate to generate knowledge. Can a particular subject position be supported by an a priori argument? It can, but only by an argument that claims a particular structure for the world and then identifies a particular subjectivity as uniquely capable of knowing that structure. The problem with such arguments is that they beg the question. The one subject position that could be advanced as epistemically superior to others without presupposing something about the structure of the world is the unconditioned position, the position of no position that provides a view from nowhere. Attractive as this ideal might seem, arguments in the philosophy of science suggest that this is a chimera. Let me turn to them.

FEMINIST EPISTEMOLOGICAL STRATEGIES 2: MULTIPLYING SUBJECTS

The ideal of the unconditioned (or universally conditioned) subject is the traditional proposal for escaping the particularity of subjectivity. Granting the truth of the claim that individual subjectivities are conditioned, unconditioned subjectivity is treated as an achievement rather than a natural endowment. The methods of the natural sciences constitute means to that achievement. . . . The difficulty just outlined for the feminist epistemological strategy of changing the subject, however, has a parallel in developments in the philosophy of science. Both dilemmas suggest the individual knower is an inappropriate focus for the purpose of understanding (and changing) science.

In the traditional view, the natural sciences are characterized by a methodology that purifies scientific knowledge of distortions produced by scientists' social and personal allegiances. The essential features of this methodology—explored in great detail by positivist philosophers of science—are observation and logic. Much philosophy of science in the last twenty-five years has been preoccupied with two potential challenges to this picture of scientific methodology—the claim of Kuhn, Feyerabend, and Hanson that observation is theory laden and the claim of Pierre Duhem that theories are underdetermined by data. One claim challenges the stability of observations themselves, the other the stability of evidential relations. Both accounts have seemed (at least to their critics and to some of their proponents) to permit the unrestrained expression of scientists' subjective preferences in the content of science. If observation is theory laden, then observation cannot serve as an independent constraint on theories, thus permitting subjective elements to constrain theory choice. Similarly, if observations acquire evidential relevance only in the context of a set of assumptions, a relevance that changes with a suitable change in assumptions, then it's not clear what protects theory choice from subjective elements hidden in background assumptions. Although empirical adequacy serves as a constraint on theory acceptance, it is not sufficient to pick out one theory from all contenders as the true theory about a domain of the natural world. These analyses of the relation between observation, data, and theory are often thought to constitute arguments against empiricism, but, like the feminist epistemological strategies, they are more effective as arguments against empiricism's silent partner, the theory of the unconditioned subject. The conclusion to be drawn from them is that what has been labelled scientific method does not succeed as a means to the attainment of unconditioned subjectivity on the part of individual knowers. And as long as the scientific knower is conceived of as an individual, knowing best when freed from external influences and attachment (that is, when detached or free from her/his context), the puzzles introduced by the theory-laden nature of observation and the dependence of evidential relations on background assumptions will remain unsolved.

It need not follow from these considerations, however, that scientific knowledge is impossible of attainment. Applying what I take to be a feminist insight—that we are all in relations of interdependence—I have suggested that scientific knowledge is constructed not by individuals applying a method to the material to be known but by individuals in interaction with one another in ways that modify their observations, theories and hypotheses, and patterns of reasoning. Thus scientific method includes more than just the complex of activities that constitutes hypothesis testing through comparison of hypothesis statements with (reports of) experiential data, in principle an activity of individuals. Hypothesis testing itself consists of more than the comparison of statements but involves equally centrally the subjection of putative data, of hypotheses, and of the background assumptions in light of which they seem to be supported by those data to varieties of conceptual and evidential scrutiny and criticism.[6] Conceptual criticism can include investigation into the internal and external consistency of a hypothesis and investigation of the factual, moral, and social implications of background assumptions; evidential criticism includes not only investigation of the quality of the data but of its organization, structuring, and so on. Because background assumptions can be and most frequently are invisible to the members of the scientific community for which they are background and because unreflective acceptance of such assumptions can come to define what it is to be a member of such a community (thus making criticism impossible), effective criticism of background assumptions requires the presence and expression of alternative points of view. This sort of account allows us to see how social values and interests can become enshrined in otherwise acceptable research programmes (i.e. research programmes that strive for empirical adequacy and engage in criticism). As long as representatives of alternative points of view are not included in the community, shared values will not be identified as shaping observation or reasoning.

Scientific knowledge, on this view, is an outcome of the critical dialogue in which individuals and groups holding different points of view engage with each other. It is constructed not by individuals but by an interactive dialogic community. A community's practice of inquiry is productive of knowledge to the extent that it facilitates transformative criticism. The constitution of the scientific community is crucial to this end as are the interrelations among its members. Community level criteria can, therefore, be invoked to discriminate among the products of scientific communities, even though context-independent standards of justification are not attainable. At least four criteria can be identified as necessary to achieve the transformative dimension of critical discourse:

1. There must be publicly recognized forums for the criticism of evidence, of methods, and of assumptions and reasoning.
2. The community must not merely tolerate dissent, but its beliefs and theories must change over time in response to the critical discourse taking place within it.
3. There must be publicly recognized standards by reference to which theories, hypotheses, and observational practices are evaluated and by appeal to which criticism is made relevant to the goals of the inquiring community. With the possible exception of empirical adequacy, there needn't be (and probably isn't) a set of standards common to all communities. The general family of standards from which those locally adopted might be drawn would include such cognitive virtues as accuracy, coherence, and breadth of scope, and such social virtues as fulfilling technical or material needs or facilitating certain kinds of interactions between a society and its material environment or among the society's members.
4. Finally, communities must be characterized by equality of intellectual authority. What consensus exists must not be the result of the exercise of political or economic power or of the exclusion of dissenting perspectives; it must be the result of critical dialogue in which all relevant perspectives are represented.

Although requiring diversity in the community, this is not a relativist position. True relativism, as I understand it, holds that there are no legitimate constraints on what counts as reasonable to believe apart from the individual's own beliefs. Equality of intellectual authority does not mean that anything goes but that everyone is regarded as equally capable of providing arguments germane to the construction of scientific knowledge. The position outlined here holds that both nature and logic impose constraints. It fails, however, to narrow reasonable belief to a single one among all contenders, in part because it does not constrain belief in a wholly unmediated way. Nevertheless, communities are constrained by the standards operating within them, and individual members of communities are further constrained by the requirement of critical interaction relative to those standards. To say that there may be irreconcilable but coherent and empirically adequate systems for accounting for some portion of the world is not to endorse relativism but to acknowledge that cognitive needs can vary and that this variation generates cognitive diversity.

. . .

273

DILEMMAS OF PLURALISM

This sort of account is subject to the following dilemma.[7] What gets produced as knowledge depends on the consensus reached in the scientific community. For knowledge to count as genuine, the community must be adequately diverse. But the development of a theoretical idea or hypothesis into something elaborate enough to be called knowledge requires a consensus. The questions must stop somewhere, at some point, so that a given theory can be developed sufficiently to be applied to concrete problems. How is scientific knowledge possible while pursuing socially constituted objectivity? That is, if objectivity requires pluralism in the community, then scientific knowledge becomes elusive, but if consensus is pursued, it will be at the cost of quieting critical oppositional positions.

My strategy for avoiding this dilemma is to detach scientific knowledge from consensus, if consensus means agreement of the entire scientific community regarding the truth or acceptability of a given theory. This strategy also means detaching knowledge from an ideal of absolute and unitary truth. I suggest that we look at the aims of inquiry (at least some) as satisfied by embracing multiple and, in some cases, incompatible theories that satisfy local standards. This detachment of knowledge from universal consensus and absolute truth can be made more palatable than it might first appear by two moves. One of these is implicit in treating science as a practice or set of practices; the other involves taking up some version of a semantic or model-theoretic theory of theories.

Beginning with the second of these, let me sketch what I take to be the relevant aspects of implications of the semantic view.[8] This view is proposed as an alternative to the view of theories as sets of propositions (whether axiomatized or not). If we take the semantic view, we understand a theory as a specification of a set of relations among objects or processes characterized in a fairly abstract way. Another characterization would be that on the semantic view, a theory is the specification of a structure. The structure as specified is neither true nor false; it is just a structure. The theoretical claim is that the structure is realized in some actual system. As Mary Hesse has shown, models are proposed as models of some real world system on the basis of an analogy between the model and the system, that is, the supposition that the model and the system share some significant features in common.[9] Models often have their start as metaphors. Examples of such metaphoric models are typical philosophers' examples like the billiard ball model of particle interactions or the solar system model of the atom. What many feminists have pointed out (or can be understood as having pointed out) is the use of elements of gender ideology and social relations as metaphors for natural processes and relations. Varieties of heterosexual marriage have served as the metaphoric basis for models of the relation between nucleus and cytoplasm in the cell, for example.[10] The master molecule approach to gene action, characterized by unidirectional control exerted on organismal processes by the gene, reflects relations of authority in the patriarchal household. Evelyn Fox Keller has recently been investigating the basis of models in molecular biology in androcentric metaphors of sexuality and procreation.[11] When Donna Haraway says that during and after the Second World War the organism changed from a factory to a cybernetic system, she can be understood as saying that the metaphor generating models of orgasmic structure and function shifted from a productive system organized by a hierarchical division of labour to a system for generating and processing information.[12] Alternatively put, cells, gene action, and organisms have been modelled as marriage, families, and factories and cybernetic networks, respectively. Supporting such analysis of particular theories or models requires not merely noticing the analogies of structure but also tracing the seepage of language and meaning from one domain to another as well as studying the uses to which the models are put.

The adequacy of a theory conceived as a model is determined by our being able to map some subset of the relations/structures posited in the model onto some portion of the experienced world. (Now the portions of the world stand in many relations to many other portions.) Any given model or schema will necessarily select among those relations. So its adequacy is not just a function of isomorphism of one of the interpretations of the theory with a portion of the world but of the fact that the relations it picks out are ones in which we are interested. A model guides our interactions with the interventions in the world. We want models that guide the interactions and interventions we seek. Given that different subcommunities within the larger scientific community may be interested in different relations or that they may be interested in objects under different descriptions, different models (that if taken as claims about an underlying reality would be incompatible) may well be equally adequate and provide knowledge, in the sense of an ability to direct our interactions and interventions, even in the absence of a general consensus as to what's important. Knowledge is not detached from knowers in a set of propositions but consists in our ability to understand the structural features of a model and to apply it to some particular portion of the world; it is knowledge of that portion of the world through its structuring by the model we use. The notion of theories as sets of propositions requires that we view the adequacy of a theory as a matter of correspondence of the objects, processes, and relations described in the propositions of the theory with the objects, processes, and relations in the domain of the natural world that the theory purports to explain; that is, it requires that adequacy be conceptualized as truth. The model-theoretic approach allows us to evaluate theories in relation to our aims as well as in relation to the model's isomorphism with elements of the modelled domain and permits the adequacy of different and incompatible models serving different and incompatible aims. Knowledge is not contemplative but active.

The second move to escape the dilemma develops some consequences of treating science as practice. There are two worth mentioning. If we understand science as practice, then we understand inquiry as ongoing, that is, we give up the idea that there is a terminus of inquiry that just is the set of truths about the world. (What LaPlace's demon knew, for example.) Scientific knowledge from this perspective is not the static end point of inquiry but a cognitive or intellectual expression of an ongoing interaction with our natural and social environments. Indeed, when we attempt to identify the goals of inquiry that organize scientific cognitive practices, it becomes clear that there are several, not all of which can be simultaneously pursued.[13] Scientific knowledge, then, is a body of diverse theories and their articulations onto the world that changes over time in response to the changing cognitive needs of those who develop and use the theories, in response to the new questions and anomalous empirical data revealed by applying theories, and in response to changes in associated theories. Both linear-reductionist and interactionist models reveal aspects of natural processes, some common to both and some uniquely describable with the terms proper to one but not both sorts of model. If we recognize the partiality of theories, as we can when we treat them as models, we can recognize pluralism in the community as one of the conditions for the continued development of scientific knowledge in this sense.

In particular, the models developed by feminists and others dissatisfied with the valuative and affective dimensions of models in use must at the very least (given that they meet the test of empirical adequacy) be recognized as both revealing the partiality of those models in use and as revealing some aspects of natural phenomena and processes that the latter conceal. These alternative models may have a variety of forms and a variety of motivations, and they need not repudiate the aim of control. We engage in scientific inquiry to direct our interactions with the interventions in the world. . . . If we aim for effective action in the natural world, something is to be controlled. The issue should be not whether but what and how. Rather than repudiate it, we can set the aim of control within the larger context of overall purposes and develop a more refined sense of the varieties of control made possible through scientific inquiry.

A second consequence for feminist and other oppositional scientists of adopting both the social knowledge thesis and a model-theoretic analysis of theories is that the constructive task does not consist in finding the one best or correct feminist model. Rather, the many models that can be generated from the different subject positions ought to be articulated and elaborated. Very few will be exclusively feminist if that means exclusively gender-based or developed only by feminists. Some will be more appropriate for some domains, others for others, and some for none. We can't know this unless models get sufficiently elaborated to be used as guides for interactions. Thus, this joint perspective implies the advocacy of subcommunities characterized by local standard. To the extent that they address a common domain and to the extent that they share some standards in common, these subcommunities must be in critical dialogue with each other as well as with those subcommunities identified with more mainstream science. The point of dialogue from this point of view is not to produce a general and universal consensus but to make possible the refinement, correction, rejection, and sharing of models. Alliances, mergers, and revisions of standards as well as of models are all possible consequences of this dialogic interaction.

CONCLUSIONS

Understanding scientific knowledge in this way supports at least two further reflections on knowledge and power. First of all, the need for models within which we can situate ourselves and the interactions we desire with the natural world will militate against the inclusiveness required for an adequate critical practice, if only

277

because the elaboration of any model requires a substantial commitment of material and intellectual resources on the part of a community.[14] This means that, in a power-stratified society, the inclusion of the less powerful and hence of models that could serve as a resource for criticism of the received wisdom in the community of science will always be a matter of conflict. At the same time, the demand for inclusiveness should not be taken to mean that every alternative view is equally deserving of attention. Discussion must be conducted in reference to public standards, standards which, as noted above, do not provide timeless criteria, but which change in response to changes in cognitive and social needs. Nevertheless, by appeal to standards adopted and legitimated through processes of public scrutiny and criticism, it is possible to set aside as irrelevant positions such as New Age 'crystalology' or creationism. To the extent that these satisfy none of the central standards operative in the scientific communities of their cultures, they indeed qualify as crackpot. Programmes for low-tech science appropriate to settings and problems in developing nations may, by contrast, be equally irritating to or against the grain of some of the institutionalized aspects of science in the industrialized nations, but as long as they do satisfy some of the central standards of those communities, then the perspectives they embody must be included in the critical knowledge-constructive dialogue. Although there is always a danger that the politically marginal will be conflated with the crackpot, one function of public and common standards is to remind us of that distinction and to help us draw it in particular cases. I do not know of any simple or formulaic solution to this problem.

Second, . . . the structures of cognitive authority themselves must change. No segment of the community, whether powerful or powerless, can claim epistemic privilege. If we can see our way to the dissolution of those structures, then we need not understand the appropriation of power in the form of cognitive authority as intrinsic to science. Nevertheless, the creation of cognitive democracy, of democratic science, is as much a matter of conflict and hope as is the creation of political democracy.

Notes

1. Empiricist philosophers have found themselves in great difficulty when confronting the necessity to make their theory of the knower explicit, a difficulty most eloquently expressed in David Hume's Appendix to *A Treatise of Human Nature*, ed. L. A. Selby-Bigge (Oxford: Clarendon Press, 1960).
2. The later philosophy of Wittgenstein does challenge the individualist ideal.

Until recently few commentators have developed the anti-individualist implications of his work. See Naomi Scheman, 'Individualism and the Objects of Psychology', in Sandra Harding and Merrill Hintikka (eds.), *Discovering Reality* (Boston: Reidel, 1983), 225–44.

3. Cf. Longino, 'Can There Be A Feminist Science?', in *Hypatia* 2/3 (Autumn 1987); and ch. 7 of Longino, *Science as Social Knowledge* (Princeton: Princeton University Press, 1990).

4. Cf. Evelyn F. Keller, 'The Force of the Pacemaker Concept in Theories of Slime Mold Aggregation', in *Perspectives in Biology and Medicine*, 26 (1983), 515–21; and *A Feeling for the Organism* (San Francisco: W. H. Freeman, 1983).

5. Evelyn F. Keller, *Reflections on Gender and Science* (New Haven: Yale University Press, 1984).

6. For argument for and exposition of these points, see Longino, *Science as Social Knowledge*, esp. ch. 4.

7. Thanks to Sandra Mitchell for this formulation.

8. My understanding of the semantic view is shaped by its presentations in Bas van Fraassen, *The Scientific Image* (New York: Oxford University Press, 1980); and Ronald Giere, *Explaining Science* (Chicago: University of Chicago Press, 1988).

9. Mary Hesse, *Models and Analogies in Science* (Notre Dame, Ind.: Notre Dame University Press, 1966).

10. The Gender and Biology Study Group, 'The Importance of Feminist Critique for Contemporary Cell Biology', in *Hypatia*, 3/1 (1988).

11. Evelyn Fox Keller, 'Making Gender Visible in the Pursuit of Nature's Secrets', in Teresa de Lauretis (ed.), *Feminist Studies/Critical Studies* (Bloomington: Indiana University Press, 1986), 67–77; and 'Gender and Science', in *The Great Ideas Today* (Chicago: Encyclopedia Britannica, 1990).

12. Donna Haraway, 'The Biological Enterprise: Sex, Mind, and Profit from Human Engineering to Sociobiology', in *Radical History Review*, 20 (1979): 206–37.

13. This point is developed further in *Science as Social Knowledge*, ch. 2.

14. For a somewhat different approach to a similar question, see Philip Kitcher, 'The Division of Cognitive Labour', in *Journal of Philosophy*, 87/1 (Jan. 1990), 5–23.

Bibliography

An enormous amount of work elaborating feminist critiques of the sciences has been published in the last fifteen years. We list here only some of the major works (monographs and anthologies) articulating theoretical dimensions of that critique as well as work in theory of language, philosophy of science, and science studies upon which feminist thinkers have drawn. Many of the volumes cited fall into more than one category. We have given the full citation in one category and indicated in brackets at the end of each section those works in other categories that could also qualify.

1. Feminism and Science

Major monographs by some of the central participants in the development of feminist theory about the sciences.

Fausto-Sterling, Anne, *Myths of Gender: Biological Theories About Women and Men* (New York: Basic Books, 1985; 2nd edn., 1992).

Haraway, Donna, *Simians, Cyborgs, and Women: the Reinvention of Nature* (New York: Routledge, 1991).

Harding, Sandra, *The Science Question in Feminism* (Ithaca, NY: Cornell University Press, 1986).

Keller, Evelyn Fox, *Reflections on Gender and Science* (New Haven: Yale University Press, 1985).

Longino, Helen, *Science as Social Knowledge: Values and Objectivity in Scientific Inquiry* (Princeton: Princeton University Press, 1990).

Merchant, Carolyn, *The Death of Nature: Women, Ecology, and the Scientific Revolution* (San Francisco: Harper & Row, 1980; London: Wildwood House, 1980).

Rose, Hilary, *Love, Power, and Knowledge: Towards a Feminist Transformation of the Sciences* (Cambridge: Polity Press, 1994).

Schiebinger, Londa L., *The Mind Has No Sex? Women in the Origins of Modern Science* (Cambridge, Mass.: Harvard University Press, 1989).

[See also Keller, in section 2; Harding, Nelson in section 3; Keller, Haraway in section 5.]

These anthologies contain some major papers:

Harding, Sandra and Jean F. O'Barr, *Sex and Scientific Inquiry* (Chicago: University of Chicago Press, 1987).

Tuana, Nancy, *Feminism and Science* (Bloomington Ind.: Indiana University Press, 1989).

[See also Harding and Hintikka (eds.) in section 3.]

For an extensive bibliography of work through 1989, see Wylie, Alison, Kathleen Okruhlik, Sandra Morton, and Leslie Thielen-Wilson,

'Philosophical Feminism: A Bibliographic Guide to Critiques of Science', *Resources for Feminist Research*, 19/2 June 1990) (New Feminist Research), which constitutes an exhaustive bibliography of work through 1989.

2. Science, Language, and Metaphor

General works about metaphor upon which feminist thinkers have drawn:

Black, Max, *Models and Metaphors: Studies in Language and Philosophy* (Ithaca, NY: Cornell University Press, 1962).

Hesse, Mary, *Models and Analogies in Science* (Notre Dame, Ind.: University of Notre Dame Press, 1966).

Kittay, Eva, *Metaphor: Its Cognitive Force and Linguistic Structure* (Oxford and New York: Oxford University Press, 1987).

Lakoff, George and Mark Johnson, *Metaphors We Live By* (Chicago: University of Chicago Press, 1980).

Ortony, Andrew (ed.), *Metaphor and Thought* (Cambridge and New York: Cambridge University Press, 1979, 1980; 2nd end. 1993).

Works on metaphor and science:

Beer, Gillian, *Darwin's Plots: Evolutionary Narrative in Darwin, George Eliot, and Nineteenth-Century Fiction* (London: Routledge & Kegan Paul, 1983).

Levine, George, *One Culture: Essays in Science and Literature* (Madison: University of Wisconsin Press, 1987).

Rotman, Brian, *Signifying Nothing: The Semiotics of Zero* (Stanford, Calif.: Stanford University Press, 1993; Basingstoke: Macmillan, 1987).

Young, Robert, *Darwin's Metaphor: Nature's Place in Victorian Culture* (Cambridge: Cambridge University Press, 1985).

Feminist analyses of scientific language:

Jordanova, L. J., *Sexual Visions: Images of Gender in Science and Medicine Between the Eighteenth and Twentieth Centuries* (Madison: University of Wisconsin Press, 1993).

Keller, Evelyn Fox, *Secrets of Life, Secrets of Death: Essays on Language, Gender, and Science* (New York: Routledge, 1992).

3. Feminist Theory of Knowledge

Major monographs in feminist theory of knowledge:

Code, Lorraine, *What Can She Know? Feminist Theory and the Construction of Knowledge* (Ithaca, NY: Cornell University Press, 1991).

Collins, Patricia Hill, *Black Feminist Thought: Knowledge, Consciousness, and the Politics of Empowerment* (New York: Routledge, 1991; London: Harper Collins Academics, 1991).

Harding, Sandra, *Whose Science? Whose Knowledge? Thinking from Women's Lives* (Ithaca, NY: Cornell University Press, 1991).

Hekman, Susan, *Gender and Knowledge: Elements of a Postmodern Feminism* (Cambridge: Polity, 1990; Boston: Northeastern University Press, 1992).

Nelson, Lynn Hankinson, *Who Knows?: From Quine to a Feminist Empiricism* (Philadelphia: Temple University Press, 1990).

Scheman, Naomi, *Engenderings: Constructions of Knowledge, Authority, and Privilege* (New York: Routledge, 1993).

Smith, Dorothy, *The Conceptual Practices of Power: a Feminist Sociology of Knowledge* (Toronto: University of Toronto Press, 1990; Boston: Northeastern University Press, 1990).

[See also Keller, Harding, Longino in section 1.]

Important anthologies:

Alcoff, Linda and Elizabeth Potter (eds.), *Feminist Epistemologies* (New York: Routledge, 1993).

Antony, Louise and Charlotte Witt (eds.), *A Mind of One's Own: Feminist Essays on Reason and Objectivity* (Boulder: Westview Press, 1993).

Haack, Susan (ed.), 'Feminist Epistemology: For and Against' (special issue: *The Monist: An International Journal of General Philosophical Inquiry*, 77/4 (Oct. 1994)).

Harding, Sandra and Merrill B. Hintikka (eds.), *Discovering Reality: Feminist Perspectives on Epistemology, Metaphysics, Methodology, and Philosophy of Science* (Dordrecht, Neth.: D. Reidel; Boston, 1983).

Jaggar, Alison M. and Susan Bordo (eds.), *Gender/Body/Knowledge: Feminist Reconstructions of Being and Knowing* (New Brunswick, NJ: Rutgers University Press, 1989).

Lennon, Kathleen and Margaret Whitford (eds.), *Knowing the Difference: Feminist Perspectives in Epistemology* (London: Routledge, 1994; New York: Routledge, 1994).

4. Philosophy of Science

Anthologies offering a good sense of the range of issues addressed in recent and contemporary philosophy of science.

Boyd, Richard, Philip Gasper, and J. D. Trout (eds.), *The Philosophy of Science* (Cambridge, Mass.: MIT Press, 1991).

Brody, Baruch A. and Richard E. Grandy, *Readings in the Philosophy of Science* (Englewood Cliffs, NJ: Prentice Hall, 2nd edn. 1989).

Philosophical texts on which feminist theorists of scientific knowledge have drawn:

Cartwright, Nancy, *How the Laws of Physics Lie* (Oxford and New York: Oxford University Press, 1983).

Giere, Ronald, *Explaining Science: a Cognitive Approach* (Chicago: University of Chicago Press, 1988).

Hacking, Ian, *Representing and Intervening: Introductory Topics in the Philosophy of Natural Science* (Cambridge and New York: Cambridge University Press, 1983).

Hesse, Mary B., *Revolutions and Reconstructions in the Philosophy of Science* (Brighton: Harvester Press, 1980; Bloomington, Ind.: Indiana University Press, 1980).

Hollis, Martin and Steven Lukes (eds.), *Rationality and Relativism* (Oxford: Blackwell, 1982; Cambridge, Mass.: MIT Press, 1st end., 1982).

Kuhn, Thomas S., *The Structure of Scientific Revolutions* (Chicago: University of Chicago Press, 1962, 2nd edn., 1970).

[See also Longino, Nelson in section 2; Hesse in section 3.]

Monographs that draw on work by feminist theorists of science:

Dupré, John, *The Disorder of Things: Metaphysical Foundations of the Disunity of Science* (Cambridge, Mass.: Harvard University Press, 1993).

Kellert, Stephen H., *In the Wake of Chaos: Unpredictable Order in Dynamical Systems* (Chicago: University of Chicago Press, 1993).

Rouse, Joseph, *Knowledge and Power: Toward a Political Philosophy of Science* (Ithaca, NY: Cornell University Press, 1987).

5. Science Studies

Monographs:

Bloor, David, *Knowledge and Social Imagery* (London and Boston: Routledge & Kegan Paul, 1976; Chicago: University of Chicago Press, 2nd edn., 1991).

Cadden, Joan, *Meanings of Sex Difference in the Middle Ages: Medicine, Science, and Culture* (Cambridge and New York: Cambridge University Press, 1993).

Haraway, Donna, *Primate Visions: Gender, Race and Nature in the World of Modern Science* (London: Verso, 1989; New York: Routledge, 1989).

Keller, Evelyn Fox, *A Feeling for the Organism: the Life and Work of Barbara McClintock* (San Francisco: W. H. Freeman, 1983).

Latour, Bruno, *The Pasteurization of France* (Cambridge, Mass.: Harvard University Press, 1988).

— and Steven Woolgar, *Laboratory Life: the Construction of Scientific Facts* (Princeton: Princeton University Press, 1986, 1979; Beverly Hills, Calif.: Sage Publications, 1979).

Porter, Theodore M., *The Rise of Statistical Thinking, 1820–1900* (Princeton: Princeton University Press, 1986).

Sapp, Jan, *Beyond the Gene: Cytoplasmic Inheritance and the Struggle for Authority in Genetics* (New York: Oxford University Press, 1987).

Shapin, Steven and Simon Schaffer, *Leviathan and the Air Pump* (Princeton: Princeton University Press, 1985).

Traweek, Sharon, *Beamtimes and Lifetimes: the World of High Energy and Physicists* (Cambridge, Mass.: Harvard University Press, 1988).

[See also Haraway and Rose in section 1; Beer, Rotman, and Keller in section 3.]

Anthologies:

Galison and Stump (eds.), *Disunities and Contextualism* (Stanford University Press, forthcoming).

Knorr-Cetina, K. and Michael J. Mulkay (eds.), *Science Observed: Perspectives on the Social Study of Science* (London and Beverly Hills, Calif.: Sage Publications, 1983).

Megill, Alan (ed.) *Rethinking Objectivity* (Durham, NC: Duke University Press, 1994).

Pickering, Andrew (ed.), *Science as Practice and Culture* (Chicago: University of Chicago Press, 1992).

Index